Packet Switching Evolution from Narrowband to Broadband ISDN

For a complete listing of the *Artech House Telecommunications Library*,
turn to the back of this book...

Packet Switching Evolution from Narrowband to Broadband ISDN

M. Smouts

Artech House
Boston • London

Library of Congress Cataloging-in-Publication Data

Smouts, M.
    Packet switching evolution from N-ISDN to B-ISDN / M. Smouts.
       p.   cm.
    Includes bibliographical references and index.
    ISBN 0-89006-542-X
    Packet switching (Data transmission) 2. Integrated services digital networks. I. Title.
    TK5105.S59 1991   91-36642
    621.382—dc20   CIP

British Library Cataloguing-in-Publication Data

Smouts, M.
    Packet switching evolution from N-ISDN to B-ISDN.
    I. Title
    004.66

ISBN 0-89006-542-X

© Artech House.
685 Canton Street
Norwood, MA 02062

All rights reserved. Printed and bound in the United States of America. No part of this book may be reproduced or utilized in any form or by any means, electronic or mechanic including photocopying, recording, or by any information storage and retrieval system, without permission in writing from the publisher.

International Standard Book Number: 0-89006-542-X
Library of Congress Catalog Card Number: 91-36642

10 9 8 7 6 5 4 3 2 1

Contents

PREFACE 1

CHAPTER 1 INTRODUCTION 3

CHAPTER 2 PACKET SWITCHING IN ISDN 13

   2.1. Basic Principles 13
   2.2. Packet-Handling Techniques 17
   2.3. Services 25
   2.4. DataLink Layer Protocols 26
   2.5. Subscriber Numbering 29
   2.6. Packet Call Scenario 32
   2.7. Speed Of Call Establishment And Packet Delay 35
   2.8. Charging Principles 37
   2.9. The ISDN Concentrator 39
   2.10. Flow Control 40
   2.11. Congestion and Overload Control 43

CHAPTER 3 EVOLUTION IN PACKET SWITCHING 47

   3.1. Comparison of Possible Solutions 47
      3.1.1. Evolution of networks 47
      3.1.2. Short Description of Packet Networks 49
   3.2. Maximum Integration of Packet Switching within ISDN Networks 54
      3.2.1. The Network 54
      3.2.2. Packet-Switching Module Configuration 56
      3.2.3. Software Structure 60

3.3.	Minimum Integration of Packet-Switching within ISDN	62
	3.3.1. The Network	62
	3.3.2. The Network Elements	63
	3.3.3. ISDN Packet-Mode Bearer Services (PMBS)	66
	3.3.4. Signaling on the Packet Handler Interface	73
3.4.	Packet-Switching Network Aspects	75
	3.4.1. Why Bother with Packet-Switching?	75
	3.4.2. Why Maximum Integration?	75
	3.4.3. Disadvantages of a Minimum Integration Network	76
	3.4.4. Material Cost Difference	78
	3.4.5. Extra Functions Needed to Convert a Minimum Integration Network into a Maximum Integration Network	79
	3.4.6. Quality of Service Comparison	80
	3.4.7. Disavantages of X.25	81
	3.4.8. Interexchange Signaling Aspects	81
	3.4.9. Conclusion	82
3.5.	User-to-User Signaling	83
	3.5.1. Packetized Signaling in ISDN Networks	83
	3.5.2. Architecture	84
	3.5.3. User–to–User Signaling Applications	91
	3.5.4. Signaling System Number 7	92
	3.5.5. User–to–User Signaling in Signaling System Number 7	105
3.6.	New Packet Mode	108
	3.6.1. Introduction	108
	3.6.2. Frame Relaying Bearer Service	109
	3.6.3. U-Plane Congestion Control (Reference CCITT I.370)	118
	3.6.4. Applicability of Frame Mode Bearer Services for ISDN	122
	3.6.5. Interworking Between NPM and Other Protocols	125
	3.6.6. Advantages of Frame Relaying	127
3.7.	MAN (Metropolitan Area Network)	129
	3.7.1. Introduction	129
	3.7.2. Other High-Speed Networks	130
	3.7.3. MAN Basic Switching Principle	130
	3.7.4. Network Structure	134
	3.7.5. Services	138
	3.7.6. Q3 and SMDS Protocol	138
	3.7.7. Product Evolution	141
3.8.	Asynchronous Transfer Mode	144

	3.8.1. How to Handle Higher Bit Rates	144
	3.8.2. The ATM Technique	147

CHAPTER 4 DESCRIPTION OF AN IMPLEMENTATION 151

- 4.1. The ISDN Chip Set 151
- 4.2. ISDN Modules 155
  - 4.2.1. **Performance Requirements** 157
  - 4.2.2. ISM 158
  - 4.2.3. IPTM (ISDN Packet Trunk Module) 164
  - 4.2.4. IRSU (ISDN Remote Subscriber Unit) 171
- 4.3. Operation and Maintenance for Data Communication 173
  - 4.3.1. Errors and Alarms 173
  - 4.3.2. Display of Status 174
  - 4.3.3. Traffic Measurement and Statistics 175
  - 4.3.4. Traffic Management 175

APPENDIX A PACKET-HANDLING BACKGROUND 177

- A.1. Packet Switching in General 177
- A.2. X.25 Packet Handling 178
  - A.2.1. X.25 Level 1 178
  - A.2.2. X.25 Level 2 179
  - A.2.3. X.25 Level 3 182
  - A.2.4. X.75 187
  - A.2.5. X.25 Conclusions 187
- A.3. Packetized Signaling 189
  - A.3.1. Signaling System Number 7 189
  - A.3.2. Digital Subscriber Signaling Number 1 190
- A.4. LAPM 192
- A.5. New Packet Mode (Additional Packet Mode Bearer Service) 192
- A.6. Asynchronous Transfer Mode (ATM) 195
  - A.6.1. ATM and HDLC 195
  - A.6.2. ATM and New Packet Mode 196
  - A.6.3. ATM and X.25 197

ABBREVIATIONS 199

INDEX 203

Preface

The best way to introduce packet switching into the ISDN (Integrated Services Digital Network) is a subject that has given rise to much controversy. This is partly because most telephone administrations have two conflicting interests: to maintain or extend the services in the public telephone network in competition with the private networks on the one hand, and to defend the services already provided by the existing separate public data network on the other hand. ISDN is an alternative public data network that extends the services of the public telephone network, but it enters into competition with the public data network, partly because the suppliers of switching equipment have a varying stake in promoting the telephone network in favor of the data network. Some have succeeded more than others in adapting their switches to the new requirements. Add to all of this the fact that ISDN is only at the start of a general takeoff, and enough reasons are found to explain why the necessary infrastructure to build a fully integrated network with packet switching services included is not yet available.

In the absence of the appropriate infrastructure, this book explains the important aspects to be considered when selecting the best way to introduce packet switching into ISDN; for instance, the compatibility of services between circuit switching and packet switching. It discusses advantages and disadvantages for subscribers and network operators and demonstrates how solutions can be offered that can evolve from the actual situation to a fully integrated network in the future. Dimensioning, performance, and cost aspects are also explained.

The best possible solution today, given the lack of standards for a new packet mode, is a combination of semipermanent links to the packet-switching network with packet handlers in the ISDN network. A nice and easy complement to this is user-to-user signaling, which serves the low-duty applications in a very flexible way. By implementing the network in this way, all the components are already in place to provide a solid basis for the next step, which is full integration of packet switching in the telephone network.

Because any possible next move will coincide with the establishment of standards for ATM (asynchronous transfer mode), care must be taken to prepare a scenario of smooth integration of wide and broadband services in this same network. Frame-relaying and cell-relaying techniques will have to be designed such that an easy transition from one to the other will be possible.

Chapter 1

Introduction

Tie-in to the Existing Environment

When designing the best possible telecommunication network one should take into account among others the following characteristics:

For call handling:

- bandwidth per channel
- efficiency for the protocols
- subscriber facilities
- delay of call setup
- quality of service (for example, lost call rate)

For operation and maintenance:

- ease of extension
- efficiency of maintenance and defense mechanisms
- flexibility of operations (for example, charges and errors reports)

All of these characteristics must be provided in a high-quality, low-cost network, allowing the telephone administrations to supply the services expected by the subscribers at an acceptable cost.

Nobody will contest the above statement, but when trying to achieve this objective one must operate with the equipment already installed in the network and choose the best possible scenario to meet the demand for new services, at the same time paving the way for additional future services.

In the telecommunication network of today, digital technology is being introduced into a network still consisting partly of electromechanical switches, while also trying to forecast which hooks have to be provided in order to cope with future broadband services. This would be a difficult enough task in the context of one supplier, one administration, and one network, but the world presents a multisupplier, multiadministration, multinetwork environment. What is very clear from a distance (during the design phase) becomes much more blurred as the day of realization comes closer.

Circuit Switching Alone or with Packet Switching

Since the start of the design of ISDN it was very clear that packet switching had to be integrated within the voice switching network. Comparing the success of packet switching data networks (in the early days of data switching) and circuit switching data networks, one can say that packet switching often leads to a more economical data network than circuit switching. Advantages of communicating data in the form of packets are
- improved line utilization by interleaving packets belonging to different calls, thus taking advantage of silent periods in individual calls
- possible procedure and speed conversion, permitting intercommunication between different types of terminals
- high-quality service, obtained by providing automatic correction of transmission errors and reconfiguration without data loss in the event of a breakdown

Therefore, when a transport mechanism has to be selected for the transportation of data, only a few applications benefit from circuit switching; packet switching should be the obvious choice for the majority.

An Alternative

The packet-switched public data network (PSPDN) is an available alternative, but can it cope with the expected growth of data traffic (because most PSPDNs are built as small networks)? If the predictions materialize the growth of the networks could be one or two orders of magnitude (see Figure 1.1). The PSPDNs would have to increase in scale, whereas the additional cost to introduce packet switching into the ISDN is just a few percents, because the resources available for circuit switching in ISDN can be used (after some minor modifications) to handle packets. In some countries a PSPDN hardly even exists. Thus, if the ISDN doesn't provide the packet-switching service, no data service is available in the public networks.

The Place of the Packet Handler

So, where should the packet handler be located?

If a particular resource is not available in its own network and the call has to be switched through to another network in order to perform the requested function, then the total amount of equipment necessary to establish the call is too great. Too much transmission equipment is needed to set up the requested connection.

The obvious case is that of a packet-switching call between two centrex subscribers belonging to the same group and both connected to the ISDN, with the packet-switching function located in the PSPDN. The call between two adjacent subscribers is switched through the local exchange and several transit exchanges, through the public data network, and then back to the local exchange, possibly passing again through several transit exchanges, although the connection could have stayed within one printed circuit board. Thus the data should be switched as close as possible to the origin.

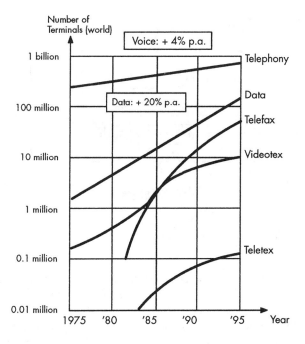

Figure 1.1 Growth of voice and data communication terminals.

The Fully Integrated Network

The basic principle of the ISDN is to provide in one network all services that were up to now provided by three different networks (see Figure 1.2). If everything indicates that ISDN should be the network to meet the demand for data services and that packet switching is the natural bearer for the transport of data, then why is the introduction of ISDN slower than expected and why is packet switching not integrated at once? Will the evolution go as shown in Figure 1.3 or will ISDN remain an access network?

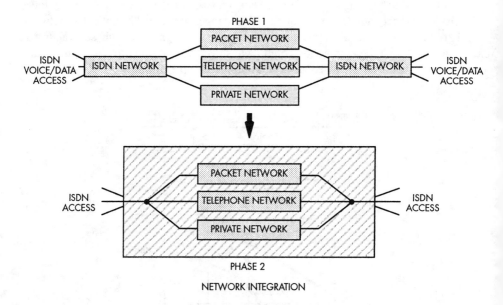

Figure 1.2 ISDN total integration.

Although the first trials of ISDN networks provided fully integrated packet switching, the commonly agreed ISDN network in Europe will give access only to the packet-switched public data network (PSPDN), and the ISDN subscriber will be a subscriber of the PSPDN for packet services. In the so-called integrated services digital network (ISDN), the service packet switching and all related subscriber facilities will not be integrated. The question to integrate or not to integrate has been answered by a temporary compromise: minimum integration. This term in itself can be, and is sometimes, understood in different ways:

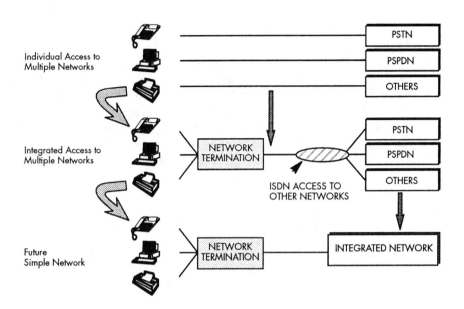

Figure 1.3 Two-step evolution of ISDN.

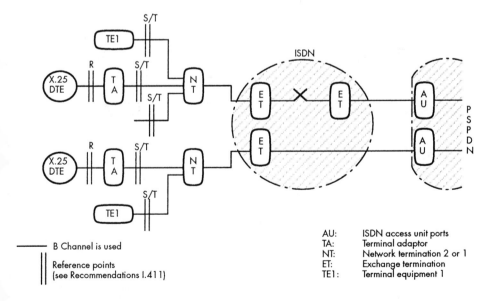

Figure 1.4 X.31 case A.

- Minimum integration refers to X.31 case A (see Figure 1.4); maximum integration refers to case B (see Figure 1.5).
- Minimum integration refers to a packet handler in the packet-switched public data network (PSPDN); maximum integration offers a packet handler in the ISDN.
- Minimum integration refers to packet services on the B channel only; maximum integration provides packet services on both B and D channels.
- Minimum integration is implemented by separating telephone and packet traffic close to the subscriber equipment, and each service is handled by its own network (see Figure 1.6); the maximum integration network handles both types of traffic.

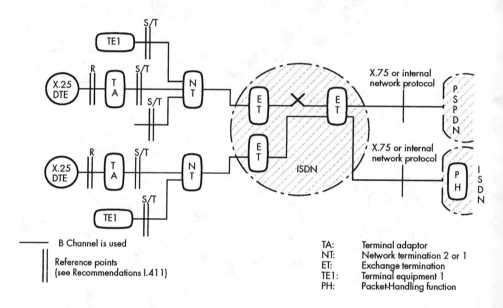

Figure 1.5 X.31 case B.

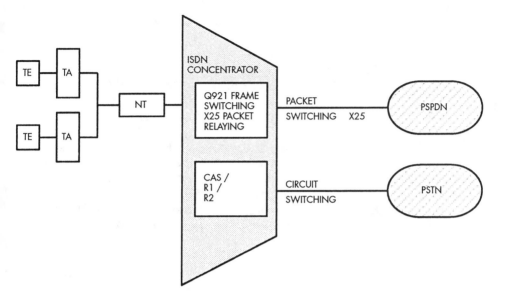

Figure 1.6 Introduction of ISDN with only concentrators.

In this book the term *minimum integration* means that ISDN provides an access to the PSPDN only (see Figure 1.7), and the switching of the packets is done by the PSPDN. The term *maximum integration* means that ISDN and PSPDN are two separate networks (see Figure 1.8). The first handles both voice and data traffic; for data there are two possibilities: circuit and packet switching. The second handles only data and only in the form of packets.

Minimum or Maximum?

If, in an international forum, a decision of this kind must be made, besides political and economical reflections most consideration is given to call-handling related functions. This means: can the basic services be supplied? Other considerations such as operation and maintenance aspects or implementation cost are more difficult to consider. Nevertheless, questions such as the following are as important as the call-handling functions:

How many subscriptions do I need?
How many bills do I get?
Am I paying two monthly fees?
How do I subscribe to, for example, call interception for a circuit-switched call or for a packet-switched call?

What is my subscriber number as a telephone call or as a packet call?
Can I use subadressing?
Can I transfer a call between two terminals?
How is fault diagnosis done in the network?
How many manipulations are necessary to assign a new or additional facility to a subscriber?
In how many places is charging done?
Where are statistics collected?

If maximum integration is selected, a few more detailed selections will have to be done as well. Is packet switching or frame switching used, or is frame relaying used? Can supplementary services be lined up between circuit switching and packet switching? Is there still any resemblance with X.25, and is X.75 used as trunk signaling? Which functions are treated in band, and which out of band? Will line and trunk signaling be lined up? Can packet switching be supported up to 2 Mb/s? How close is ATM to narrowband ISDN packet switching?

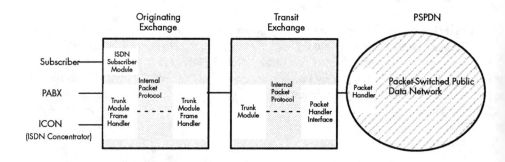

Figure 1.7 Minimum integration.

Thirty-Nine Scotts Road Singapore 0922 Telephone: 7376888
Telex: RS 37750 SHNSIN Telefax: 7371072

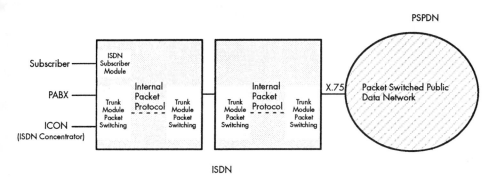

Figure 1.8 Maximum integration.

While standardization work that will give general answers to these questions is going on, we will discuss a few of the consequences. If minimum integration is the fact of life today, which benefits do we detect in it? We can consider it as an enhancement for the PSPDN subscribers because their network will be extended by more users. Their network accessibility is improved because the ISDN network may have a higher penetration and will bring the network closer to the user. Furthermore, with the use of X.25 on the D channel, the cost of access through the ISDN to the PSPDN is very competitive with the use of modems.

As long as only bilateral standards exist for establishing a data call over the international borders, the PSPDN network can serve this purpose in a more generalized way. Any gaps left by the minimum integration can be complemented by user-to-user signaling, which provides, as will be explained later, a very straightforward data path between two applications that would otherwise have to use the more complicated X.25 signaling. Examples of user-to-user applications are centrex, signaling between two PABXs over the public switching network, smart phones, and directory inquiries. So, are we prepared for the expected increase in data traffic of 20 percent per year? Can the current situation help us to draw some conclusions that will facilitate the introduction of broadband functions in the telecommunication networks in an easier way than ISDN?

The intention of this book is to contribute in a positive way to the decision making. The answers to all the questions cannot be given on the basis of technical considerations alone, but if the technical picture is clear, any economical or political reflections become easier also.

Chapter 2

Packet Switching in ISDN

2.1. BASIC PRINCIPLES

The Network Access

In the ISDN network each user has access to the network through a network termination (NT), which is a box installed on the user's premises that allows for three channels: B1, B2, and D, of 64 kb/s, 64 kb/s, and 16 kb/s, respectively. Thus a total of 144 kb/s useful bitrate is transmitted between the subscriber and the exchange (see Figure 2.1). A subscriber can use the two B channels for either circuit-switching or packet-switching calls. The

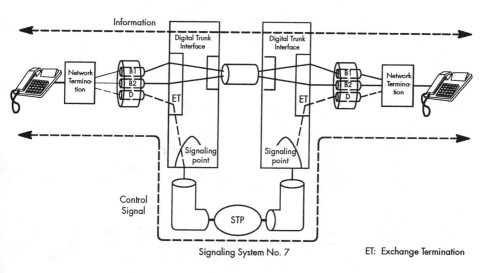

Figure 2.1 Subscriber access and circuit-switched connection.

D channel is used for transmitting subscriber line signaling information but is not fully utilized because signaling packets occupy the D channel for only 5 percent of the available capacity. The rest of the capacity can be used for packet switching over the D channel.

Differences Between Circuit and Packet Switching

When a call is set up using circuit switching, a connection is established in the form of a full duplex synchronous data path. The D channel is used for the signaling information transmission between the subscriber and the exchange, and the signaling between exchanges follows the SS7 signaling network. Once the data path is established, the users can continuously and freely exchange data. The connection can be represented as one solid pipe, because only one physical layer is utilized (see Figure 2.2).

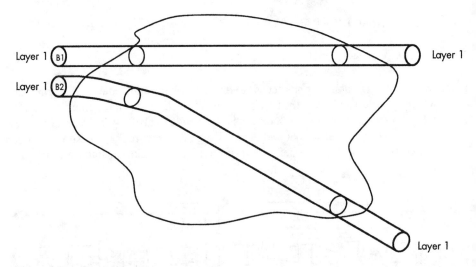

Figure 2.2 Circuit-switching network.

If a packet-switched call is set up, no solid path needs to be allocated between two parties. Each party transmits information in the form of packets that are individually transported to the other party over the network. Depending on the facilities available in the network the call is set up using circuit-switching principles to reach a point in the network at which a packet-switching function (or packet handler) is available. From there on, when several switching nodes are present in the network, X.75 signaling can be used as trunk signaling between exchanges (see Figure 2.3). Packets are sent over a path without repeating the called party address. Bursts of information are sent from one side to the other, and the pauses between packets are of varying durations, depending on the volume of traffic being transmitted. The network has to ensure that packets are received in the correct sequence.

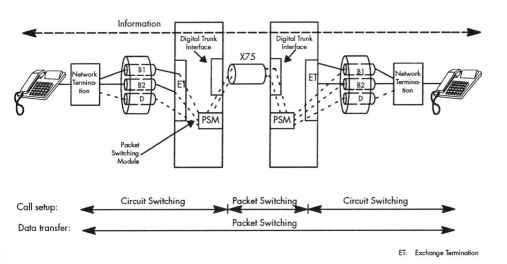

Figure 2.3 Packet switching in ISDN.

Whereas a circuit connection fully uses the physical channel, several packet connections can share the same channel. This can be achieved by all transmitting at the maximum channel speed (64 kb/s or 16 kb/s), but not continuously. The network will take care of interleaving the packets of different users. This can also be done by multiplexing packets, sent by the users at a lower bitrate than the maximum bitrate of the channel, for example, 2,4 kb/s. In the case of multiplexing, each subscriber gets its own separate channel of a lower speed.

Another possibility that packet switching can provide and circuit switching cannot is the communication between one party and several other parties, for example, a computer with several terminals. The network will take care of the necessary rate adaptations. If the cost rates are favorable, a full network of terminals connected to a host computer could be permanently established, the network taking care of reestablishment in case of failure. Charging would be proportional to the amount of data sent or received. In this case circuit-switched call would require a set of semipermanent connections. Even if the full bitrate (64 kb/s) is not required continuously, billing is done as if it were used all the time (because nobody else can employ it).

Connectionless Packet Switching

A special case, not supported by ISDN today, is the connectionless packet-switching option. This corresponds to a very short call sending only one packet. Indeed, each packet contains the necessary routing information to find its way through the network, and each packet is routed through the network independent from the others.

X.25

One form of packet protocol is X.25, in which case the subscriber-network interface is structured in three layers (see Figure 2.4):

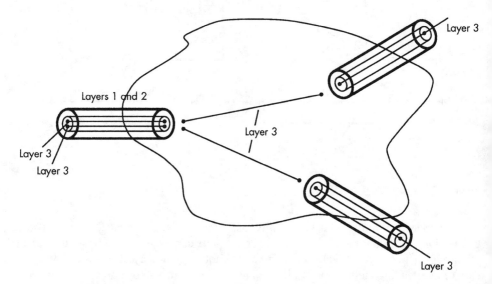

Figure 2.4 Packet switching network.

- Layer 1 is the physical transmission channel which handles electrical interface, sequencing, circuit identification, and so on. It supports one layer 2 connection.
- Layer 2 takes care of sequencing, redundancy, error detection, frame or packet retransmission, synchronization, multiplexing of packets, and quality of service parameters.
- Layer 3 uses layer 2 functions and provides support for signaling and data transmission by exercising the necessary network connections, flow control, release, and so on.

Because a B channel can be used in both circuit-switching and packet-switching modes, at the beginning of the call a preliminary signaling phase is required on the D channel in order to set the B channel into packet mode. From then on X.25 in band signaling can be used on the B channel. Because a D channel can perform only packet-switching functions and not circuit-switching functions, the call can be immediately set up in packet-switching mode.

Packet mode protocols other than X.25 have been defined, such as system network architecture (SNA), which tries to provide a more flexible signaling system out of band instead of in band, simplify the layered structure and line up the services in the network between the connection modes (for example, between circuit and packet switching).

2.2. PACKET-HANDLING TECHNIQUES

Depending on the degree of control exercised by the network on the data sent between two users, different techniques are recognized. First, several principles have to be understood. The following definitions may vary between documents from different organizations, so care must taken to go back to the definitions each time.

A *frame* is a string of bits consisting of several fields: an opening and closing flag, a field for redundancy bits (called a *frame check sequence*) and an information field (see Figure 2.5). In the latter the actual user data or call setup data are found; this is the actual packet. The data sent or received in the information field belongs to layer 3. The information in the other fields belongs to layer 2.

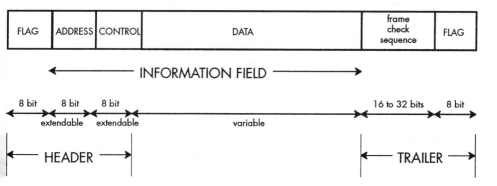

Figure 2.5 Frame format.

A *packet* is either call setup information during the setup phase of the call or the real data that two subscribers in conversation want to exchange between each other. A packet is transmitted in an envelope called a *frame*. As noted above, a packet within the frame is surrounded by information that is used for layer 2 functions (sequencing, error detection, and so on). A diagramatic representation of the relation between layers and frames is shown in Figure 2.6. For an example of a packet transmitted during call setup (for example, a call-accepted packet) see Figure 2.7. An example of a data packet is depicted in Figure 2.8.

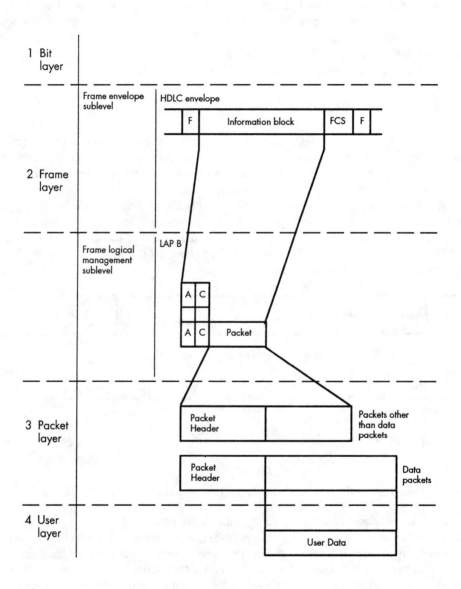

Figure 2.6 Relationship between logical levels and frame content.

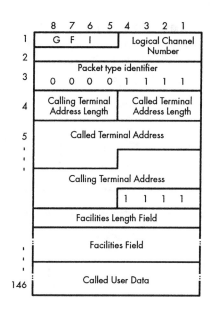

Figure 2.7 Call-accepted packet format.

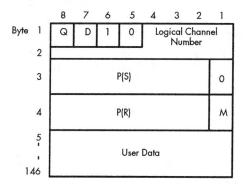

Q bit: is the data qualifying bit that the subscriber can use to distinguish between two types of data, for example, control data and actual data
D bit: used for subscribers, end-to-end acknowledgment, that is, for confirmation that a packet has been received
P(S): the packet sequence number in the send (S) direction
P(R): the packet sequence number in the receive (R) direction
M bit: the MORE bit, which means that a packet is transmitted in several parts and is split over several frames

Figure 2.8 Data packet format.

Packet switching (see Figure 2.9) is defined in X.25. Flow control and congestion control is effected between each two nodes of the network at both layers 2 and 3. Routing of packets and treatment of packet loss is handled at layer 3. Some redundancy between layers 2 and 3 is unavoidable.

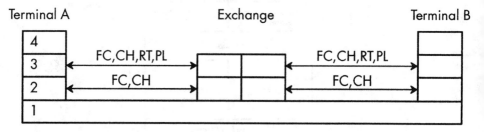

FC: Flow control
RT: Routing
CH: Congestion handling
PL: Packet loss treatment

Figure 2.9 Packet switching.

In *packet relaying* (see Figure 2.10), no flow control is executed at layer 3, only routing of packets. This is the only difference between packet switching and packet relaying. In fact packet relaying means that the link between two users is not interrupted in the network in a buffering function: the packets arrive at the receiver at the same speed and cadence as they were sent out by the sender. If layer 2 doesn't succeed at treating the virtual calls on the same connection on an equal basis, the probability that one user can block another is high.

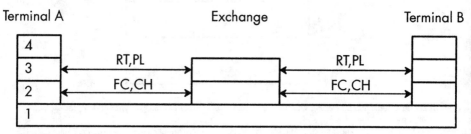

FC: Flow control
RT: Routing
CH: Congestion handling
PL: Packet loss treatment

Figure 2.10 Packet relaying.

Frame switching (see Figure 2.11) differs from packet switching in that it doesn't analyze a frame down to the level of the packet: there is no layer 3 control. Flow control is done at layer 2 at the same place routing is done, so big users cannot disturb small users. For call setup layer 3 functions are necessary. Therefore, either out-of-band signaling is used for call setup, or if in-band signaling is used, the header of the frame has to distinguish between call setup and conversation phase, and then during the call setup phase the frame content has to be analyzed deeper by the network. During conversation the layer 3 functions are left to the terminals at both sides.

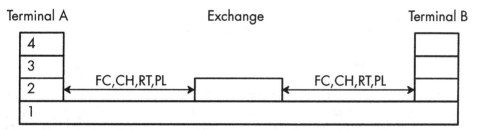

FC: Flow control
RT: Routing
CH: Congestion handling
PL: Packet loss treatment

Figure 2.11 Frame switching.

In frame switching, the network terminates the link at a point at which information can be temporarily stored in order to perform speed adaptation between the two extremes (that is, flow control). *Frame relaying* (see Figure 2.12) is different from frame switching in that it does not provide this buffer point in the network for speed adaptation. No flow control is provided by the network, this is left to the terminals at both sides. Frame relaying will be applied only in high-quality transmission networks, such as those based on optical fiber links.

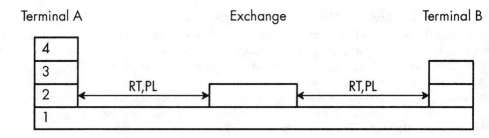

RT: Routing
PL: Packet loss treatment

Figure 2.12 Frame relaying.

The content of the frames and their sequence is still guaranteed but frames lost by the network cannot be recovered; retransmission must be requested by the receiving terminal. Variants of frame relaying include type 1 (see Figure 2.13), with no lost frame recovery, and type 2, which foresees an edge–to–edge lost frame recovery (see Figure 2.14). Combinations of frame switching and frame relaying are possible, such as frame switching type 1 (see Figure 2.15) and frame switching type 2 (see Figure 2.16).

Cell relaying is a variant of frame relaying type 2 because end–to–end lost frame recovery is specified (see Figure 2.17.). Cell relaying, in contrast to frame relaying, uses fixed–length frames; these frames are then called *cells*. Both asynchronous transfer mode (ATM) and metropolitan area network (MAN) use cell relay technology. As in frame relaying, cell relaying also minimizes the amount of processing by the network and assumes high-quality transmission links to transfer the data.

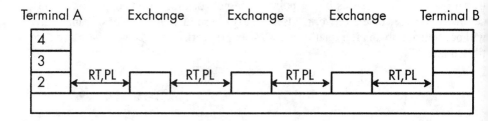

RT: Routing
PL: Packet loss treatment

Figure 2.13 Frame relaying type 1.

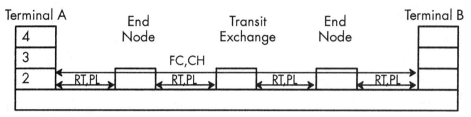

FC: Flow control
RT: Routing
CH: Congestion handling
PL: Packet loss treatment

Figure 2.14 Frame relaying type 2.

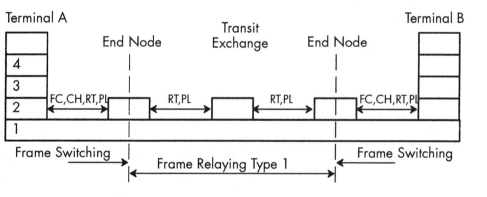

FC: Flow control
RT: Routing
CH: Congestion handling
PL: Packet loss treatment

Figure 2.15 Frame switching type 1.

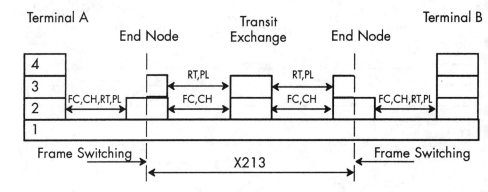

FC: Flow control
RT: Routing
CH: Congestion handling
PL: Packet loss treatment

Figure 2.16 Frame switching type 2.

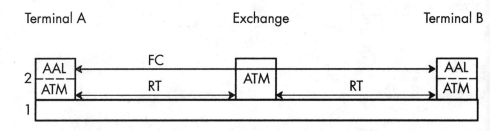

AAL: ATM adaptation layer
FC: Flow control
RT: Routing

Figure 2.17 Cell relaying.

2.3. SERVICES

A user is interested not in the technology of the network or in the internal network operation but in the service that the network can provide at a reasonable cost. The ISDN network provides three types of services (see Figure 2.18):

1. *bearer services,* which can be circuit mode or packet mode;
2. *teleservices,* which includes telephony, facsimile, teleaction, videotex and telex;
3. *supplementary services,* which includes a long list of services such as closed user groups, call transfer, identification of called or calling line, and so on.

Most supplementary services that are related to the subscriber or to the call do not differ functionally between circuit and packet switching. Both can have closed user groups, hunt groups, line identification, call redirection, restrictions on the type of call, reverse charging, call priority, and so on. However, in the case of circuit switching the network is providing only a transparent pipe (if no interworking unit is foreseen), and the quality of data transmission is left to the two terminals at both ends. In the case of packet switching, a whole range of attributes can be built into in the network:

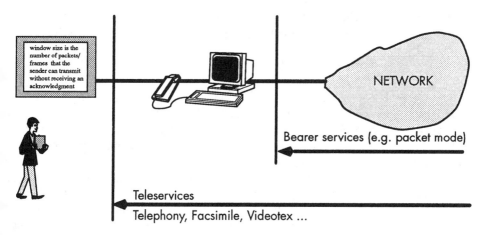

Figure 2.18 Services.

- packet length supervision
- packet size negotiation
- throughput class negotiation (that is, definition of the speed of data transmission)
- transit delay negotiation
- packet sequence numbering
- window size negotiation (the window size is the number of packets or frames that the sender can transmit without receiving an acknowledgment from the receiver)
- packet retransmission
- setup of point-to-multipoint paths for example, from several terminals to one host computer
- error correction (the correction of bit errors in case they were detected by the redundancy mechanisms)
- buffering (storing) of packet or frames
- flow control (the technique used to match the speed between sender and receiver, implemented by introducing windows)
- congestion control, executed by warning previous nodes in the network that packet traffic has to be decreased during a certain (variable) duration

2.4. DATALINK LAYER PROTOCOLS

The ISDN needs two link-access (layer 2) protocols to establish a link between two users. A link is not yet a full call; it is a switched transmission medium that, at certain points in the network, contains circuits to control errors, sequence of data elements, and so on.

LAPD

The first part of the link has to be setup with circuit-switching techniques. The protocol to be used, therefore, has been specially designed for ISDN and is called LAPD (link access protocol D). This protocol is specified by CCITT in recommendation Q.921. It uses the D channel to send information and is based on the LAPB protocol in terms of principles used, frame structure, and so on (see further below).

The subscriber's D channel is used for several purposes. Each basic access (BA) can be connected to up to eight physical terminals. These terminals can be telephones, circuit-switching devices, or packet-switching devices. The signaling information sent by those terminals is multiplexed on the D channel. The LAPD protocol has to ensure that those different types of signaling can be distinguished by the exchange. To do this the LAPD employs a two–part address consisting of a terminal endpoint identifier (TEI), which is the address that indicates which of the up to eight physical terminals (or more than eight if the eight physical terminals represent more logical terminals) has initiated the call;

and the service access point identifier (SAPI), which indicates the type of traffic that is used. The TEI and SAPI are found in the address field of the frame (see Figure 2.19).

The C/R bit in the address field identifies the frame as going in the transmit direction (bit = 0) or in the receive direction (bit = 1). The SAPI can have 64 different values, but not all are used. Those defined are as follows :
- 0 for call-control procedures, in which case the information in the frame contains signaling data to setup the call
- 1 for packet mode communication using Q.931 call-control procedures (for example, frame mode bearer service, or FMBS)
- 16 for packet mode communication using X.25 layer 3 call-control procedures
- 63 for management data, in which case the information in the frame contains data that is used by the network for operation and maintenance purposes

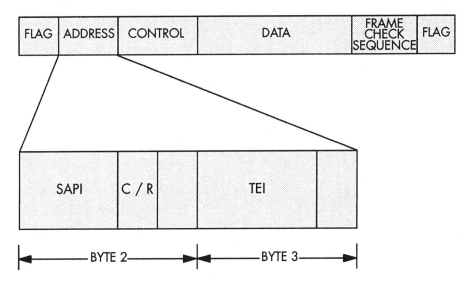

Figure 2.19 Address field of an LAPD frame.

The TEI can have 128 values. Number 127 is reserved for broadcast datalink connections, the other values are for point-to-point connections. The values 0 - 63 can be allocated by the user, the values 64 - 126 are allocated by the network.

LAPB

After the link between the subscriber and the packet-switching module is established by using the LAPD protocol, the X.25 call set up can begin. For this the LAPB (link access protocol balanced) is used. This is a layer 2 protocol that runs over the B channel to setup the call. The call setup consists of setting up the link, which is achieved by means of the LAPB. Thereafter the conversation will be further set up using layer 3. The LAPB is defined in the X.25 recommendation and was first used in PSPDNs. ISDN also makes use of it to establish the X.25 data link. The exchange of protocol information is done (see Figure 2.20) immediately between the user and the packet-switching module (PSM).

The user is the DTE (data terminal equipment) and the PSM is the DCE (data circuit terminating equipment) in the X.25 world. The setup of an X.25 call is done in the same way in the PSPDN and in the ISDN. In the ISDN the circuit-switched connection between DTE and DCE (that is, between terminal and PSM) is used as a transparent pipe.

Figure 2.20 illustrates the use of LAPD and LAPB in the different steps of link setup between terminal A and B. (A complete call scenario is explained in section 2.6.)

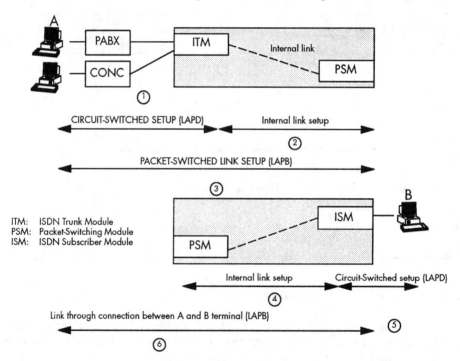

Figure 2.20 Steps involved in a call setup.

Note: In Figure 2.20 the packet-switching function is present in the local exchange. In that case the name "PSM" is used. The case in which the packet-switching function is located in the PSPDN will be considered in following chapters. In that case the name "PH" (packet handler) will be used.

1. Setup of the link between the subscriber and the ITM is completed.
2. Internal link is set up in the exchange between the ITM and the PSM.
3. A packet-switched link is set up between the A subscriber and the PSM.
4. Using the information received in the PSM about the identity of the B terminal, the PSM sets up an internal link to the ISM.
5. The ISM sets up a circuit-switched connection to the B subscriber.
6. The packet-switched link is concluded between the A and B terminals.

Difference Between LAPD and LAPB

There are two important differences between LAPD and LAPB. First LAPD has to distinguish among different types of calls – voice, circuit, packet – which it does with the SAPI as explained above. The LAPB, on the other hand, treats only X.25 calls and therefore needs only one byte in the address field. This one byte is used only to indicate the direction of the information flow; that is, the frame contains commands transferred from DTE to DCE, the frame contains responses from DTE to DCE, and so on. Second, LAPD allows multiplexing in the datalink layer using the TEI, whereas LAPB allows only a single datalink. Otherwise the structure of the frames in LAPB and LAPD are very similar. Both protocols describe how the field in the frame structure (as shown in Figure 2.19) may be filled up and how address and control fields can be used to set up a datalink, disconnect the datalink, control the information transfer, agree on retransmission of data, and so on.

There can, however, be differences in the way LAPD and LAPB are used. Imagine an ISDN network without packet switching on the D channel. The function of LAPD stops when a call of any of the different possible types is set up; LAPB takes over to control the packet flow during the data conversation after having served to set up the X.25 datalink. Then, at the release of the call, LAPD is used again to clear the connection. If D channel packet switching is used, then LAPD will control the packet flow during data conversation over the D channel. Furthermore, only LAPD is used for frame relaying and frame switching, as explained in section 3.6.

2.5. SUBSCRIBER NUMBERING

Subscriber numbering depends heavily upon the national situation, therefore what follows should be considered as an example. If the ISDN were a totally separated network, one directory number would be enough for each subscriber, even if the subscriber has various terminals connected and uses several teleservices, such as fax, teletex, videotex, and telephony. The different terminals connected to one line can be reached or identified by a subaddress, which is a set of two or more extra digits that can be added to the directory number. The teleservices are identified in the high-layer compatibility (HLC) information element, which is a set of a maximum of five bytes exchanged between two subscribers (or one subscriber and an interworking unit) to allow the called subscriber to check whether his terminal equipment is compatible with that of the calling subscriber.

However, ISDN subscribers do not call only each other, they also need to communicate with subscribers in the telephone and the data networks.

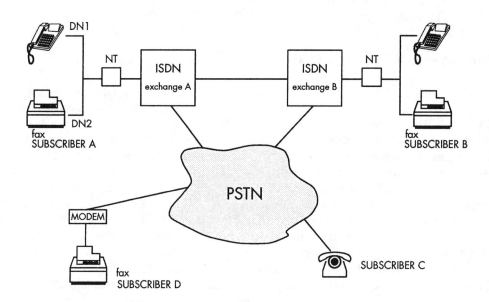

Figure 2.21 Call between PSTN and ISDN subscribers.

Interworking Between ISDN and PSTN.

If the numbering system of a pure ISDN network were used, this would mean that the already existing numbering plan would have to be changed drastically; for this reason many countries have chosen to assign another directory number (DN) to every terminal connected to the line. For an originating call the ISDN subscriber identifies himself by one of his DNs according to the service used. For a terminating call to the ISDN subscriber, the directory number dialed by the PSTN subscriber is translated by the ISDN exchange into a single DN plus the corresponding HLC element. Figure 2.21 illustrates an ISDN–PSTN interworking example. In Figure 2.21, subscriber C is connected to subscriber A by dialing DN1. The ISDN exchange A will convert this DN1 number into the ISDN directory number, which in this case is DN1 plus the HLC code signifying the teleservice telephony. This DN1 + HLC is sent through to subscriber A. A call setup by subscriber D from his telefax equipment to subscriber A will use DN2 as a called directory number. The ISDN exchange A will translate this into DN1 + HLC. The used HLC element will now mean telefax. DN1 + HLC is sent through to subscriber A to reach his telefax equipment. To set up a call to subscriber B, subscriber A dials the directory number of B and includes an HLC element to indicate the chosen teleservice.

Interworking Between ISDN and PSPDN

The subscribers of the PSPDN have a choice among several possibilities to indicate the requested terminal and teleservice: dial the DN of the wanted teleservice, transmit an HLC element to the ISDN subscriber in the user data, or fill in a subaddress in the dialed DN. An explanation about the meaning of a DN is appropriate here because the meaning of a DN is not exactly the same for ISDN and PSTN at one end and for PSPDN at the other. The numbering plan of the PSPDN follows CCITT recommendation X.121. This describes a 14-digit numbering system composed of a 4-digit data network identification code (DNIC) and a 10-digit national terminal number (NTN). The DNIC contains one digit for identifying the data network, because a country can have several separate data networks, for example, one for circuit switching and one for packet switching. The values 8, 9, and 0 of this digit are used as escape codes, meaning that when the value of this digit is, for example, 9, the number that follows identifies a number in the ISDN network and not in the PSPDN. The other three digits of the DNIC identify the country. The 10-digit NTN is the number that is dialed by another terminal within the same country and within the same data network. This number is the one found in the national directory. The NTN could further be split up into a 3-digit area code, a 3-digit central office code, or a 4-digit end point number.

The numbering plan of the PSTN and ISDN follows recommendation E.164, which specifies a variable length address of up to 15 digits. The number is composed of a country code (CC), a national destination code (NDC), and a subscriber number (SN), all of which are variable in length. An escape code is dialed in front of this 15-digit number to address a terminal outside of the ISDN network, for example, a 0 or 9 for access to the PSPDN.

Figure 2.22 depicts the numbering system for calls in the two directions. The called and calling numbers as present in the call request packet are shown. Other network interworking situations are possible; they are described in CCITT Recommendations X.122 and E.166.

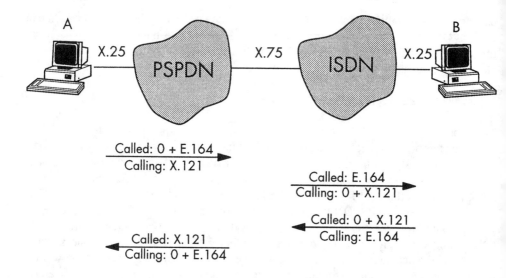

Figure 2.22 Numbering system for calls in the two directions.

2.6. PACKET CALL SCENARIO

The X.31 maximum integration scenario is chosen here as a simple example because it is fully independent of packet-switched public data networks and hence effectively integrates packet-switching within ISDN.

Both B and D channel packet switching are supported by the same access method. A local call between, for example, a basic access user (2B + D) with a D channel packet service subscription and a primary rate access user (30B + D) with B channel packet-switching capability uses the modules shown in Figure 2.23. The PABX multiplexes packet calls at layer 3 from its extension lines into channels of the primary rate access (PRA).

The X.31 access method first establishes an ISDN signaling connection, using the CCITT Q.921 (LAPD) and Q.931 procedures, between the user and the network packet-handling resource, in this case a PSM. This signaling allocates a packet bearer on either a B or D channel. Next, an X.25 call is established, and data is transferred on the established packet bearer. Figure 2.24 illustrates the various packet call phases and the function split according to OSI structured models. The numbers on the figure refer to the following description.

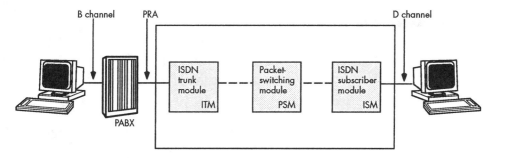

Figure 2.23 Modules necessary to set up a packet call.

1. A primary rate access user who wants to establish an X.25 virtual call first establishes a B channel connection to a packet handler in the exchange. This is indicated in the bearer capability information element of the setup message. If another virtual call has to be setup later, the Q.931 signaling phase is skipped (because the B channel has already been established), and the new X.25 call is multiplexed at layer 3 onto the existing connection.
2. ISDN call validation, primarily class and facility checks, is performed by the ITM, and a connection is then established with the packet handler using the internal exchange communication. It is assumed in the diagram that the ISM includes both circuit- and packet-handler functions, whereas in the case of primary rate access, for which traffic can be substantially higher than on a basic access, the ITM uses a separate packet handler (PSM).
3. An LAPB connection is established between user and PSM, after which the user sends an X.25 call request packet containing the ISDN directory number (E.164 numbering plan) of the destination ISDN subscriber. X.25 call validation, consisting mainly of directory number translation, class and facility checks, and charging control, is performed by the PSM.
4. Based on the result of the X.25 number translation and the terminating X.25 subscriber validation (including closed user group checks), the destination module—in this case the ISM—is contacted using the internal packet protocol.
5. At the terminating module, the necessary ISDN validation and compatibility checks are performed. If no packet bearer has been established to the B party, the ISM starts a Q.931 terminating call to acquire a D channel packet bearer. Depending on the loading algorithm used (that is, the fact that the user or the network defines whether subsequent calls can be accepted on an existing bearer), the ISM can present all terminating calls to the user using the Q.931 procedures. It is then up to the user to accept the call or to assign a new bearer. The X.25 call is multiplexed on the indicated bearer.

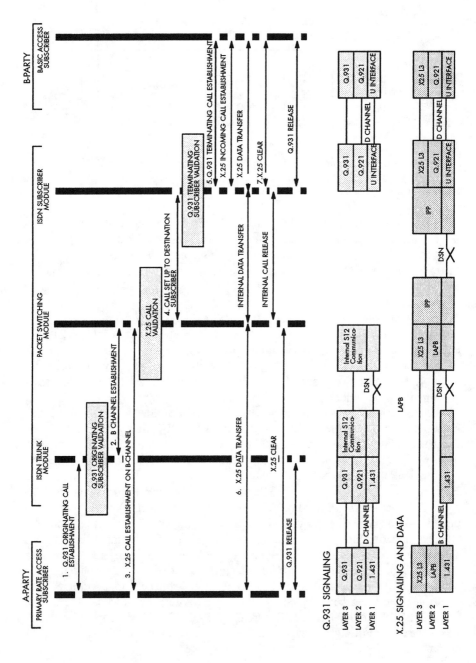

Figure 2.24 Packet call scenario.

6. Data transfer is executed using two store–and–forward points, the PSM and ISM. X.25 in–band signaling events (such as volume charging, throughput class negotiation, and reset and restart) are treated by these modules.
7. X.25 calls are released individually. The originating B channel (see point 1) or the terminating D channel packet bearer are released locally as soon as calls on them have disappeared.

2.7. SPEED OF CALL ESTABLISHMENT AND PACKET DELAY

In many existing packet-switched public data networks (PSPDNs) the subscribers are used to a small network consisting of only a few nodes (see Figure 2.25). The call setup delay in each node is typically 100 msec. (including overhead such as billing, screening, and database routing), and the time required to pass a packet from input to output of the node (packet delay) once the call is setup is around 20 msec. per node (under ideal conditions that is: no queuing). So, in a three-node PSPDN the call setup delay remains less than 300 msec. on average, and the total packet delay including transmission is less than 100 msec. (under ideal conditions).

Within the ISDN network, data calls are expected to be shorter than telephone calls. This is especially true because the available transmission speed is 64 kb/s instead of the more commonly used 2.4 kb/s in the PSPDN. The call setup delay should be short in comparison to an expected average call duration of around 30 seconds. In the existing telephone networks, call setup delays of 5 seconds are no exception, and 10 seconds is not the maximum when electromechanical exchanges are still present. The ISDN is built upon digital switches only and uses the faster signaling systems CCITT N7 for circuit switching, or X.75 for packet-switching calls.

A call is set up in typically less than 300 msec. per node whether it is for a telephone call or a circuit-switching or packet-switching call. (With the latest technology this can be improved to 200 msec.) Packet delays are around 20 msec. for each time the packet has to be stored. This can be zero times for circuit-switched calls or once or twice per node for packet-switched calls. Because the ISDN network topology is the same as for the telephone network, a maximum of 5 nodes can be expected in one call. So the total call setup delay can be, on average, 1.5 to 2 seconds, and the total packet delay can be from a few milliseconds for circuit-switched calls (the time needed for transmission only) to around 150 to 200 msec. for packet-switched calls.

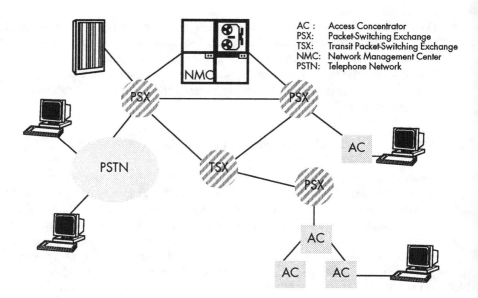

Figure 2.25 Packet-switched data network.

In the case of circuit switching a new call has to be set up every time. In the case of packet switching a call can be set up once and packets are sent over it when required. Other calls use the same connection. (This concept is called a *virtual connection*.) Of course this can be done only if the two end points of the virtual call remain the same. When using a virtual connection the call setup delay is zero.

Importance of the Delay

Narrowband data communications can be classified in one of the following three categories:

1. *dialogue-oriented*: messages of approximately half a page are exchanged between two sides, two persons, or two machines
2. *bursty traffic*: accesses to libraries with a short request in one direction and a long response
3. *interactive database access*: a short dialogue (for example, searching in a menu system and then transmission of a file), which could be modified at the other end and then sent back

The ISDN call setup delay is acceptable for all these applications. It may be high for the exceptional ones consisting of dialogue in which each monologue is as short as a few seconds. The packet delay is not important in a high-quality network. It becomes relevant when the number of retransmissions to be done amounts to several percents or in the above case of the short monologues.

2.8. CHARGING PRINCIPLES

Access Fee

A telephone line is charged for the access, which is a one-time charge for the installation of the line and the telephone. In most countries a fixed fee for renting the equipment is included in the monthly or yearly bill. A data communication is also charged for the access, but in data networks the subscriber has a choice among different types of connections:

- of speeds from 1.2 kb/s to 64 kb/s
- synchronous or asynchronous
- fixed or dial–in.

The access charge depends on all three characteristics. The monthly (or yearly) rental of the equipment also varies with the characteristics of the line. In the ISDN network most administrations have derived the charges to be applied from those valid in the telephone network. The connection fee for the two B channels and one D channel was by some administrations valued at 2.5 times that for a telephone line. Promotion of ISDN will cause these fees to go down. Dial–in connections are not used, but all other characteristics of data lines are possible. Because a B channel can be used for data as well as for voice, logically the connection fee is made similar to that for a telephone line connection.

Usage Fee

A telephone call is charged depending on the duration of the conversation and the distance between the two subscribers. In a data communication, the element of volume is used instead of duration, and volume is calculated by taking the time and speed of data transfer into account. The distance between two terminals does not influence cost except in international calls. Reductions could be given for high volume or for usage outside busy hours.

In the ISDN network both types of charging are applied: for voice calls the principles of the telephone network are used, and for data calls the principles of the data network are used. There exists, however, one difficulty in the ISDN network, which stands in the way of an all-out promotion of ISDN: the charge for distance. Although in data networks the cost of the network also increases with distance, the available bandwidth is used very efficiently, and by multiplexing channels or interleaving calls the cost of transmission is reduced drastically. In the ISDN network, where voice or circuit-switched calls need 64 kb/s for the duration of the conversation, distance has to be taken into account. Therefore, packet-switched calls in ISDN are also penalized by a distance-dependent charge. But new charging principles are being considered to bring the mechanisms closer to the ones followed by the existing data networks.

Facilities Fee

Facilities can be charged for the registration or with a monthly fee. It is also possible to charge every time the facility is used. This charge could be dependent on the complexity of the facility; for example, abbreviated dialing could be more expensive than closed user group.

Total Cost

The total cost of a connection is represented in Figure 2.26.

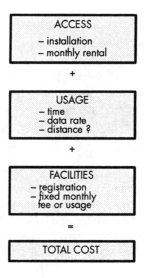

Figure 2.26 Charging cost elements.

2.9. THE ISDN CONCENTRATOR

Another key aspect of the introduction of ISDN is the strategy of using concentrators. The distance that can be covered with the two–wire connection through which the ISDN basic access is provided is limited. However, ISDN concentrators can be installed at any distance from their parent exchange because they are connected via PCM systems. This solution allows the ISDN service to be brought to an area much wider than a limited circle around the ISDN exchanges. In this way a whole country can be covered while the required number of exchanges equipped with the special functionality of ISDN can remain low.

This approach has side effects that need to be considered carefully. ISDN subscribers connected to a concentrator can be physically located in an area other than the parent exchange of this concentrator. It is required that the ISDN subscribers are assigned directory numbers consistent with their location, in line with the existing numbering plan. This means, for example, that "local" calls involving an ISDN subscriber may in fact have to be treated as interzonal calls although no area prefix has been dialed. To simplify the recognition of these calls, the following solution can be adopted: if a concentrator is connected to a host exchange in another area, then the first two digits of the local number, say, "92," are special if this combination is not yet used in any zone (see Figure 2.27).

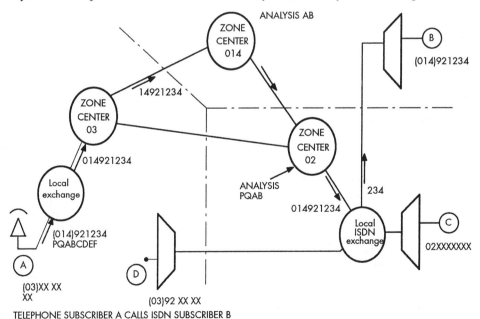

Figure 2.27 Routing to ISDN subscriber.

The charging scheme also has to be worked out to conform to this strategy of hiding the internal structure of the network to the users. As a consequence, the number of possible combinations within the same ISDN exchange is somewhat larger than is usually the case. An ISDN concentrator handles about 128 basic accesses and concentrates their traffic to one or two 2 Mb/s links, which can use a proprietary protocol or a primary rate access (PRA) protocol (see Figure 2.28). If the number of BAs to be treated is as low as 14, then the traffic doesn't have to be concentrated, and then we talk about an ISDN multiplexer. All the other functions are the same as for a concentrator. Under normal conditions the concentrator does not switch calls locally. Instead, they all go to the host exchange. B channels carrying packets are circuit-switched to a packet handler in the host exchange. A functional split between the host exchange and the concentrator could be as follows: line classes are stored in the host exchange whereas directory-number-to-equipment-number translation, device and call states, device state–dependent facilities, suspend-resume, call waiting, and point-to-multipoint are handled in the concentrator. For certain features (such as hunting), limited line class information is needed in the concentrator.

Figure 2.28 Flow control in both directions in a virtual circuit.

2.10. FLOW CONTROL

General Mechanisms

The capacity of the virtual circuit (that is, the quantity of data that can be transmitted in a given time) is limited among other things by the capabilities of the receiving DTE (for instance, the speed of its connecting line). It is therefore necessary that the transmission of data by the network to the receiving DTE be controlled by the receiver itself. For this purpose the receiving DTE sends transmission authorization to the network. This flow control is passed on by the network to the transmitting DTE in the form of transmission authorizations. (Figure 2.29 illustrates flow control.) This is done by sequence numbers.

1. Send sequence number P(S). Each data packet transmitted is numbered sequentially, this number is called the *send sequence number P(S)*. The numbering is performed modulo 8 (numbers cycle from range 0 to 7).

Note: As at frame level, relations such as "is one unit higher than," "does not exceed," and the like, are to be taken, for everything relating to sequence numbers, as modulo 8, in which $7 + 1 = 0$.

2. Receive sequence number P(R). Transmission authorization is achieved by using a receive sequence number P(R), indicating that the end (DTE or network) sending the P(R) authorizes the transmission of a further packet or packets.
3. Send window. By transmitting a P(R), the DTE or network indicates that it is ready to receive packets of number P(R) (say, 5), and the window size (say, until 7). The standard value for W, as defined in the CCITT Recommendations, is 2. Other values of W can be chosen at subscription time. In Figure 2.30, W was selected to be 3.

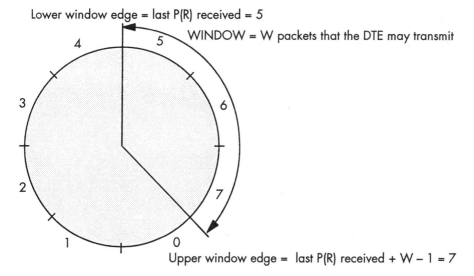

Figure 2.29 The send window.

4. Conditions for sending data packets. The transmitter (DTE or network) is authorized to transmit only packets whose P(S) is such that P(S) = last P(S) transmitted + 1 or P(S) < last P(R) received + W. The first packet transmitted for each direction of transmission, after setting up or resetting the virtual circuit, has 0 as its number. Under the same conditions, as long as no P(R) has been received, P(S) may not attain W.
5. Conditions for sending a P(R). The DTE or the network can transmit a P(R) value only if it is at least equal to the last P(R) transmitted and does not exceed the last P(S) received + 1. After setting up or resetting the virtual circuit only the value 0 for P(R) should be transmitted and must remain so until a data packet is received and

additional authorization is given.

Figure 2.30 gives a description of the window. Each packet is represented by a portion of a cycle, in which the P(S) number of the packet is indicated. It is assumed that the DTE has chosen W = 3 and that the last P(R) received from the network is equal to 5. Flow control is thus ensured independently on each virtual circuit and for each direction of transmission. The receive not ready (RNR) packet permits the receiver to notify the transmitter that for the moment it does not wish to receive any further packets. The reject (REJ) packet permits the receiver to notify the transmitter to retransmit the data packets from the indicated P(R) on.

Sequence Number Transport

The P(R)s are transported in the following ways :

1. in receive ready (RR) packets

	8 7 6 5	4 3 2 1
OCTET 1	0 0 0 1	Logical Channel Number
2		
3	P(R)	0 0 0 1

2. in receive not ready (RNR) packets

	8 7 6 5	4 3 2 1
OCTET 1	0 0 0 1	Logical Channel Number
2		
3	P(R)	0 0 1 0 1

3. in reject packets

	8 7 6 5	4 3 2 1
OCTET 1	0 0 0 1	Logical Channel Number
2		
3	P(R)	0 1 0 0 1

The P(R)s are also carried in octet 1 of the data packets circulating in the other direction of transmission (piggy-backing). The P(S) appears in octet 3 of each data packet. the following diagram shows an example of flow control from the DTE to the network (see Figure 2.30).

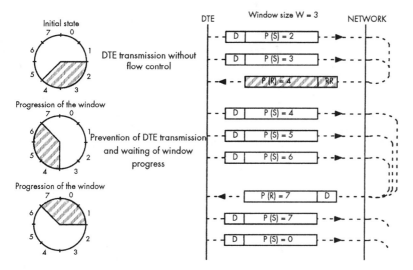

Figure 2.30 Example of flow control from DTE to network.

2.11. CONGESTION AND OVERLOAD CONTROL

In a telecommunication network for voice or circuit switching congestion in conversation phase is impossible under fault-free conditions. The user disposes of a fixed bandwidth during the total time of the conversation, whether it is employed or not.

In the packet-switching networks, resources are used in common by different subscribers in the conversation phase. If calls are multiplexed on the same channel but each data link has its own bandwidth reserved—say, 6 subscribers with a data rate of 9.6 kb/s multiplexed on the same 64 kb/s—then congestion is still not possible. In this case we talk about *static multiplexing*. If, however, several calls with the same data rate—say, 9.6 kb/s—are multiplexed in a time interleaving manner on one D channel of 16 kb/s, then congestion will arise when several data sources are sending at the same time. In order to avoid loss of information in this temporary overload situation the network will have to apply overload-control or congestion-control techniques. Voice or circuit-switched networks also use these techniques, but they do so during the call setup phase.

First of all, the best technique to avoid congestion is proper dimensioning. If the maximum load to be expected can be calculated, then congestion can be avoided by providing the necessary resources for the maximum load. Unfortunately, this may turn out to be too expensive, particularly because the load of a telecommunication network is difficult to foresee because it depends on uncontrollable elements (such as storm and war).

Furthermore, load is often not spread over the whole network, but concentrated in a restricted area. Therefore, techniques such as network management and alternate routing were put in place (see Figure 2.31). If very short overload situations are encountered, a technique that can also help is queuing (or buffering) of information. For instance, this method is used for calls to traffic operators. While waiting for call to be offered to an operator, the subscriber can listen to a recorded announcement. Buffering techniques are also applied within an exchange (for example, in a distributed control system in order to temporarily store messages sent from one processor to another). In order to avoid overload in this case the buffer has to be emptied as fast as possible, but if the receiving processor cannot follow, then at a certain threshold level (level 1; see Figure 2.32) a message is sent back to hold the traffic for a given time period t1. If the traffic flow is so abrupt that a threshold level 2 (level 2 in Figure 2.32) is reached, the traffic is interrupted for a longer time t2.

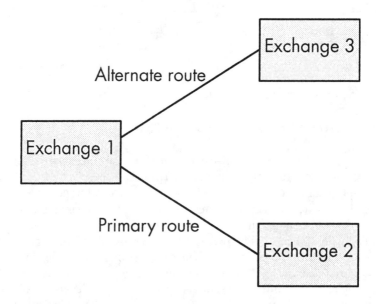

Figure 2.31 Alternative routing.

These techniques—proper dimensioning, temporarily providing extra routes, buffering, and temporarily suspending traffic—are also applied to avoid congestion in packet-switching networks. How they are applied and when and where in the network depends on the type of traffic to be treated. As said above, low-speed traffic, in which each user has reserved a bandwidth, doesn't need control. When the bandwidth becomes scarcer in proportion to the demand for it, congestion control becomes necessary.

The very high-speed applications, which are very bursty, need congestion control most. A type of protection that becomes necessary when the channel speed is higher than the one subscribed to is a policing function. If the rates are able to cope with all traffic cases a subscriber should be allowed to use all of the available bandwidth. This, however, applies only as long as the subscriber cannot block other subscribers from using the common resources. To treat very high-data rates, frame relaying is normally used (or ATM, which is a similar technique). In order to speed up the network as much as possible many functions are omitted from the network and are left to the control of the terminals at either end. One of these functions is flow control. But when the network no longer has any possibilities to control the flow of traffic it cannot avoid congestion, and that will slow down the network further than the gain that was achieved by removing flow control. Congestion control will be further discussed in section 3.6.3.

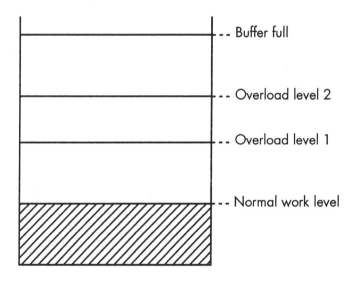

Figure 2.32 Overload levels.

Chapter 3

Evolution in Packet Switching

3.1. COMPARISON OF POSSIBLE SOLUTIONS

3.1.1. Evolution of Networks

It is not easy to predict in which direction our data networks will evolve, particularly because no other field of our modern life can serve as a good analogy to telecommunications' own typical characteristics. Nevertheless, we have to look far enough ahead to make the right decisions today.

The arrival of narrowband ISDN has added the dimension of services to telecommunications and has increased the speed of information transfer by one order of magnitude.

One more dimension–that of video–will be added by broadband ISDN, and in terms of speed of communications the plans are to improve by four orders of magnitude. In attempting to look into the next five years, which inputs will guide the flow of thought?
- What does the transition from POTS to N–ISDN teach us?
- Can ISDN replace everything?
- Which optical fiber loop will be made available?
- How far must the existing network evolve before broadband is widely introduced?
- How fast will broadband penetrate, and which services will it provide?
- How will the market be split among competing products: CATV network, extended narrowband, MANS, ATM?

- Which charging principles should be used, and how do they compare to the actual cost of features?
- What is the influence of the new networks of today: digital cellular mobile radio, intelligent networks (IN), telecommunication management network (TMN)?

The selected scenario for introduction of new techniques will have to take into account the relative success of each of the previously introduced functions. In Figure 3.1 an unqualified picture is shown of a possible future penetration of each of the networks to be evaluated. Nobody knows whether this diagram represents a real scenario, but if it is right then the necessary products have to made available to cope with this demand.

Telecommunication is a slowly moving world when one considers the average life of systems still in service. Systems invented 70 years ago are still in service and have to be interfaced with. In any modern country in the world it may take 20 more years before all rotary, cross–bar, and electromechanical exchanges will be replaced and the whole country will be digital. Functions available only in the digital systems will have to wait roughly 25 years between introduction in the field and availability to all the subscribers.

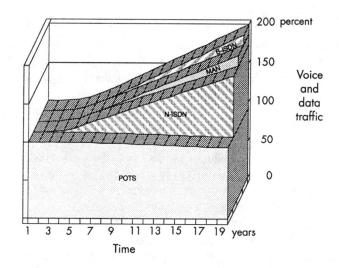

Figure 3.1 Evolution of voice and data traffic.

Of course a new service can be made available to all subscribers who want it by installing an overlay network that brings the new functions to all parts of the country, but as long as the penetration is not high, technology does not push the service, and it does not promote itself. Moreover, a new function needs two to three years to be standardized and another two to three years to be designed, tested, and introduced in the first exchanges. Care has to be taken when plotting the scenario of the future that this slowly moving context is contemplated. In the automotive and television industries it takes about ten years to replace all products belonging to a given technology. In the telecommunication industry, if a large change has to be achieved in a period of ten years, only a small part of the network can be converted because the whole infrastructure has to be adapted as well. If only a part of the technology has to be changed shorter lead times can be achieved.

In the case of introducing new packet-switching techniques, only a limited set of functions are added to the network. If, however, the scope is extended to broadband ISDN, then a whole network, covering aspects of voice, data, video and also the related operation and maintenance functions, is being considered. Therefore, it is rather unrealistic to assume that all the standards being issued by CCITT and other standardization organizations will result in viable products within the next five years. These standards include

- X.31.
- New packet mode, also called additional packet mode bearer service (APMBS), or frame relaying, or frame switching.
- Asynchronous transfer mode (ATM).
- Switched multimegabit data service (SMDS).

Of course, the application fields of these protocols are not completely overlapping.

What are the main characteristics of these different packet-switching principles in comparison to those of the basic standard X.25?

3.1.2. Short Description of Packet Networks

3.1.2.1. PSPDN

For the PSPDN the standards X.25 and X.75 are used as shown in Figure 3.2. Packets are switched by analyzing the logical channel number (LCN). Call-handling information is transmitted in band in layer 3. During data transfer, the data is acknowledged on each link: between subscriber and exchange and between each two exchanges of the PSPDN. Bitrates of up to 64 kb/s (or higher such as 128 kb/s) are supported.

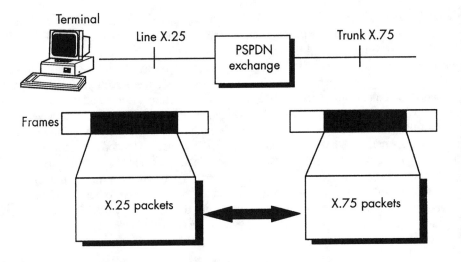

Figure 3.2 Packet-switching principle (X.25-X.75).

3.1.2.2. X.31 Maximum Integration

Packet mode services are introduced in ISDN in a pragmatic way. No new protocols were invented, but CCITT Recommendation X.31 prescribes the use of existing X.25 principles and protocols. Therefore, the existing X.25 terminal equipment can also be used. The packet-switching function is located in the ISDN (see Figure 3.3). The call is set up using the D channel signaling capabilities as for circuit switching. Packets can be sent over the D or B channel. X.75 is used for interexchange signaling. Data transfer in conversation phase is supervised as for X.25.

3.1.2.3. X.31 Minimum Integration

X.31 also allows a variant in which the switching of the X.25 call is left to PSPDN (see Figure 3.4). The ISDN network only provides the access to the PSPDN, and the user is a subscriber of the ISDN as well as of the PSPDN. The call is set up first to a packet handler located either in a central point in the ISDN network or in the PSPDN. This packet handler then expects that a message interchange is started between itself and the subscriber as for the set up of an X.25 call in the PSPDN. Frames can be interleaved on the same B channel or multiplexing techniques can be used to split one physical channel of 64 kb/s into several channels of a lower bandwidth, but the packets are routed unchanged to the packet-switching network, where the content of the packet is analyzed to execute the switching function.

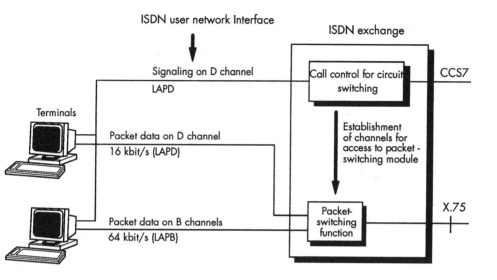

Figure 3.3 Principles of maximum integration.

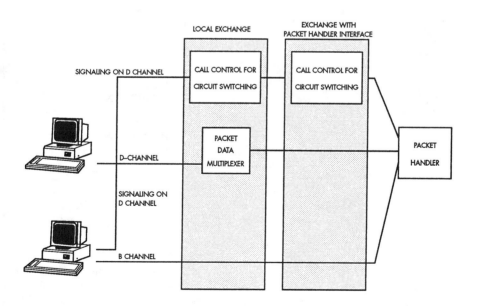

Figure 3.4 Principle of minimum integration.

3.1.2.4. New Packet Mode

No standards are available yet to define this new packet mode bearer service; however, the basic concepts have been laid down by CCITT. In this packet network the principle of out-of-band signaling as used with ISDN circuit switching is applied (see Figure 3.5). The call control information is sent by the subscriber over the D channel in layer 3, and a modified version of CCITT N7 is used between exchanges. Data are transferred on the basis of frames, that is using only the level 2 content of the frame. Several variants are still possible in which frame switching, frame relaying principles, or a combination of both, are used. The bitrate of packets to be transferred is not restricted to 64 kb/s but can go up to 30X64 kb/s or 1920 kb/s.

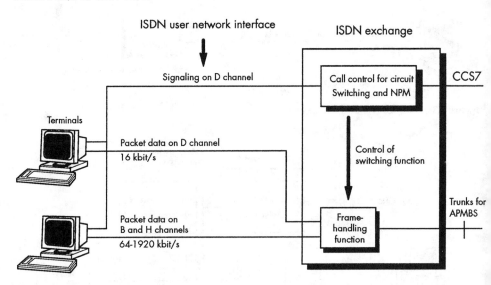

Figure 3.5 New packet mode principles.

3.1.2.5. Metropolitan Area Networks (MAN)

Some types of data communications such as the very bursty ones, can make full use of connectionless packet transmission such as the ones defined by Bellcore as switched multimegabit data service (SMDS) or by IEEE as distributed queue dual bus (DQDB). These standards are being used to interconnect LANs (local area networks) over MANs (see Figure 3.6). Connectionless packet switching does not setup a path but sends data and address in the same packet.

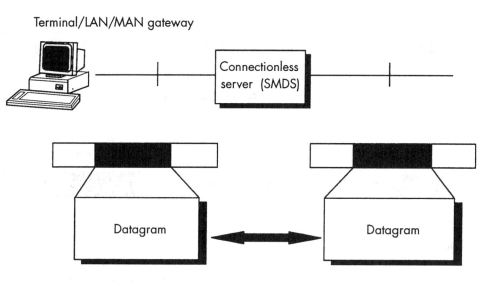

Figure 3.6 Connectionless service (SMDS).

3.1.2.6. *Asynchronous Transfer Mode (ATM)*

ATM is a principle using cell relaying for the transfer of data and is generally agreed to be the standard for broadband ISDN services. However, the concept is not really related to the switched bandwidth. It can be applied for data transmission at speeds between 64 kb/s and 2 Mb/s and it doesn't have to be transmitted over links of 600 Mb/s. The existing 2 Mb/s network can also be used.

ATM cells are switched based on the virtual channel identifier in the cell header (see Figure 3.7). The call is setup by out-of-band signaling in a virtual D channel between subscriber and exchange and an enhanced CCITT N7 signaling between exchanges. Thus again the same principles as those applied for circuit switching in ISDN are valid. The ATM concept assumes high-quality transmission media and does not provide link-by-link acknowledgment. End–to–end control is defined for ATM in order to allow users to perform possibly required segmentation—reassembly, acknowledgment, retransmission, and so on.

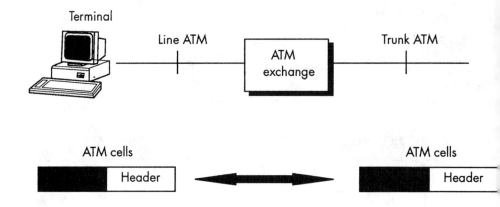

Figure 3.7 ATM switching principle.

3.2. MAXIMUM INTEGRATION OF PACKET SWITCHING WITHIN ISDN

3.2.1. The Network

Figure 3.8 shows the modules in a local exchange that are involved in integrated packet handling in one of the many possible configurations of a local exchange. The ISM (ISDN subscriber module) supports local basic accesses and can handle data packets on B and D channels. Subscriber packet protocols are terminated in the ISM. The ISM sends the packet to other modules using an internal packet protocol.

The ITM (ISDN trunk module) handles a primary rate access. It cannot terminate a subscriber packet protocol. Consequently, B channels used in the packet mode are transparently switched to a PSM (packet switching module) with subscriber packet capabilities.

The ICON (ISDN concentrator) supports remote basic accesses. It is connected via one or two standard primary rate accesses to the ITM. Data packets on basic access B or D channels are multiplexed onto a few primary rate access B channels. The other channels are used for voice or circuit-switched data.

In trunk modules, packet connections to other exchanges or to a packet network are provided by specific 64 kbit/s channels in a 2 Mbit/s digital trunk connected via a DTM (digital trunk module). The channels of the trunk module cannot handle packets, so packet-mode 64 kbit/s channels are transparently switched between the PSM and the trunk module.

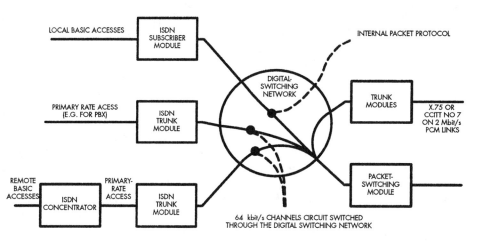

Figure 3.8 Exchange with packet-switching functions.

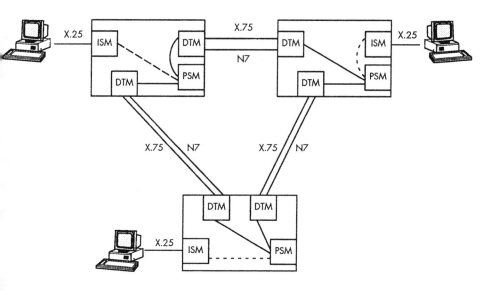

Figure 3.9 A packet-switching network within ISDN.

The packet-switching module (PSM) provides a set of packet handlers that can be reached from the ISM and the ITM for X.25 packet switching and from the trunk modules for X.75 packet switching. In Figure 3.9 shows a network with three exchanges. X.75 links are provided for packet-switching signaling between exchanges. CCITT N7 links are used for call setup of voice or circuit-switched data calls.

3.2.2. Packet-Switching Module Configuration

A possible configuration of a PSM is shown in Figure 3.10. One module consists of a processor and memory performing the layer 3 functions, a set of protocol boards performing the layer 1 and 2 functions, and interconnecting logic on the common board and the network interface board. The protocol boards allow a modem to be connected directly to the PSM or via a digital 2 Mb/s link through the exchange network (see Figure 3.11). The modem types can be V28 and V11 for speeds ranging from 2.4 kb/s to 64 kb/s. When data packets are received through the modem interface, the modem will decode the frequencies and send the resulting bits to the protocol board.

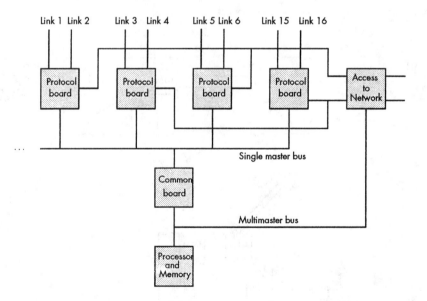

Figure 3.10 Configuration of the packet switching module.

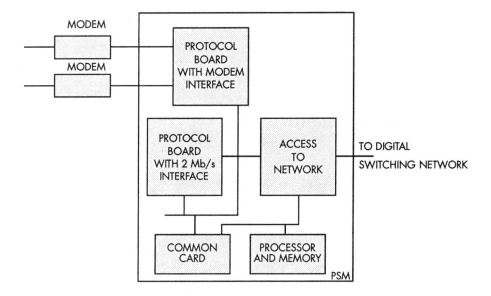

Figure 3.11 PSM interconnections.

The protocol board with the 2 Mb/s interface can handle two channels of 64 kb/s, which are connected through the network to either a trunk module or a subscriber module. Each protocol board, whether connected to a modem or to a 2 Mb/s link, contains two packet handlers for treatment of layer 1 and 2 functions (see Figure 3.12). Layer 3 is treated by the processor in the module.

When a data packet is received from the modem or the exchange network, it is first received by the protocol board in the receive queue. The necessary X.25 layer 2 functions, —sequence number validation, layer 2 flow control, retransmission, and so on— are the carried out by the protocol board. The common board is alerted to the presence of a data packet by an interrupt sent by the protocol board. When the interrupt is received, the common board reads the data packet from the protocol board memory and writes it into the RAM of the processor. The software then performs the necessary X.25 packet–switching protocol layer 3 functions.

When a data packet is to be transmitted, the software of the processor queues the data packet in the memory and alerts the common board by generating an interrupt. The common board takes control of the high–speed bus, which operates in the multimaster mode, and analyzes the data packet. The common board writes the data packet into the memory of the relevant protocol board and alerts that protocol board by generating an interrupt. The protocol board performs X.25 layer 2 functions on the data packet and transmits it to the modem or exchange network. A functional representation is depicted in Figure 3.13.

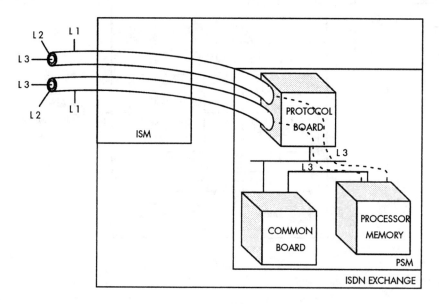

Figure 3.12 PSM treatment of layers.

3.2.2.1. Protocol Board Functions

The following tasks are performed within the protocol board :

1. Layer 1 handling, which provides modem control and maintenance in the case of external connections and channel selection and maintenance in the case of internal connections.
2. Memory, which receives data frames and transfers them to the layer 2 handler. It also informs the layer 2 handler about the completion of a frame being transmitted.
3. Layer 2 handling, which constitutes the main part of the protocol board; it implements the following X.25 and X.75 layer 2 procedures:

   - transmission, retransmission, and reception of frames, in which the maximum number of bits in the useful data content are controlled
   - layer 2 sequence number generation and checking

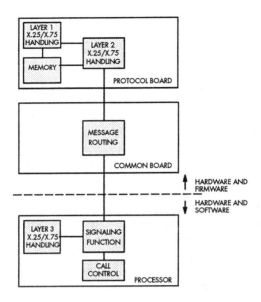

Figure 3.13 Functional representation of the PSM.

- layer 2 flow control and windowing
- LAPB state control
- memory initialization for frame transmission

In addition, the layer 2 task provides line state control and interface to the routing task in the common board and to the X.25-X.75 software in the processor for the following tasks:

- line initialization
- link setup and disconnect messages
- information frames
- flow control messages
- status request and report messages
- test request and report messages

3.2.2.2. Processor Board Functions

The processor board contains the following software functions:

- Packet-switching call control

- device handlers
- X.25 protocol handler
- X.75 protocol handler
- local charge generation
- packet-link resource manager

3.2.3. Software Structure

Layer 1 (physical layer) and layer 2 (datalink layer) functions for both X.25 and X.75 are provided by the hardware and firmware of the PSM. The layer 3 (packet or network layer) functions are provided by the packet-switching software. The ISDN packet-switching software, together with the hardware and the firmware of the PSM, provides both the X.25 DCE and the X.75 signaling terminal (STE) functions, with the software determining whether, for a given link, the PSM operates as a DCE or an STE.

Figure 3.13 shows the main functions performed by the PSM firmware and its interface with the packet-switching software. The figure also shows the division of the firmware functions between the printed board assemblies (PBAs) of the PSM. The packet-switching software performs the following main layer 3 functions:

- multiplexing of the data streams sent and received on the external links and internal connections
- control of the data flow for each data stream and control of the data flow between transmit and receive DTEs and DCEs
- detection of errors in the data streams
- reinitialization (reset, restart, and clearance) of X.25 and X.75 paths if serious errors occur

To perform the above layer 3 functions the software has to manipulate the following elements:

- logical channel identifiers (LCI) on the external links
- send and receive sequence
- numbers for packets
- control packets to regulate the flow of data
- data packets for the transfer of data
- control packets that, when necessary, reinitialize communication paths

The software structure involved in a packet-switched call in an ISDN exchange is represented in Figure 3.14. A description of each of the functional building blocks can be found in Chapter 4.

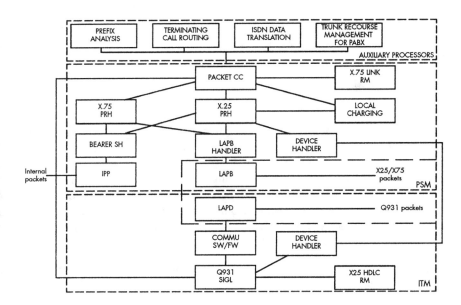

Figure 3.14 Software structure.

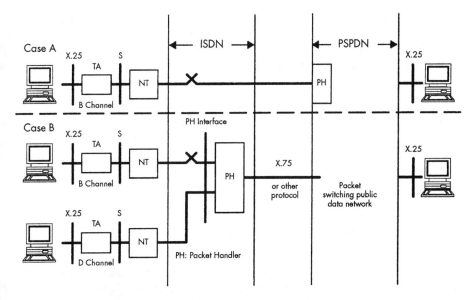

Figure 3.15 Packet mode bearer services X31 (CCITT Recommendations).

3.3. MINIMUM INTEGRATION OF PACKET-SWITCHING WITHIN ISDN

3.3.1. The Network

After the first trials of ISDN most European countries have put networks in service that are in some cases based on maximum integration packet switching. However these national networks deviate from each other and do not allow international ISDN services. Therefore, a "Memorandum of Understanding" was signed between most European countries to provide a Pan=European ISDN service. Telephone and data administrations have agreed to provide packet switching in a minimum integration manner, according to ETSI recommendation ETS 300 007 (see Figure 3.15). Two cases are defined:

1. Case A specifies only B channels and puts the packet handler (PH) in the packet-switching network.
2. Case B specifies both B and D channels and puts the packet handler at the border of the ISDN network, close to the PSPDN. This means that in all cases the packet-switching function is provided by the PSPDN, and the ISDN is used only to provide access to the packet-switching function. For B channels this access can be semipermanent (set up under control of an operator) or on demand, in which a circuit-switched connection is setup between the user and the PH. Only one datalink can be set up per channel (see Figure 3.16); each datalink can carry several calls. No multiplexation of B channels is done.

For D channels this access can also be semipermanent or on demand, as for the B channel. However, several datalinks can be multiplexed on the same D channel (see Figure 3.17). Multiplexing for D channels can be done at the following points in the network:

- in the ISDN subscriber module (ISM)
- in the ISDN concentrator
- at the outgoing point of the local exchange

At each of the points in the network at which multiplexing is done, frame relaying (or frame switching if that is selected as a better solution) has to be exercised. No multiplexing is done in the transit exchange for either D or B channels.

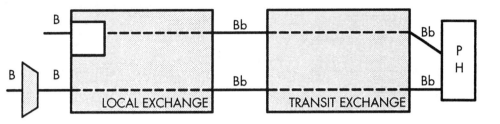

Figure 3.16 Connecting B channels to the packet handler (PH).

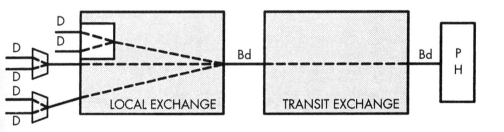

Figure 3.17 Connecting multiplexed D channels to the packet handler (PH).

3.3.2. The Network Elements

3.3.2.1. The Layers

In the minimum integration scenario the access of the subscriber is the same as in the maximum integration scenario. A packet terminal is connected to one of the two B channels (64 kb/s) or to the one D channel (16 kb/s). It can be connected in the exchange to an ISDN subscriber module to a concentrator or remote subscriber unit for remote access or to an ISDN PABX (see Figure 3.18). The B channel can be used by only one terminal at a time, the D channel by up to eight terminals at a time.

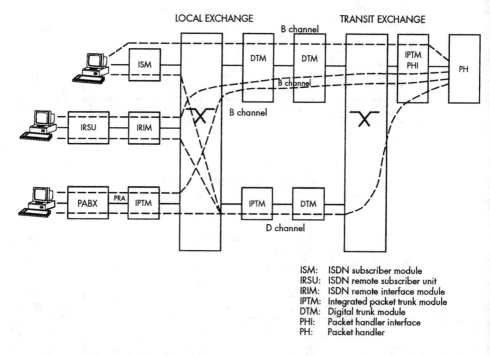

Figure 3.18 Minimum integration network elements.

Both B and D channels can be used to carry packet traffic at layer 1. Channel D must also carry the signaling information for the call setup. The physical layers (layer 1) of the B channels and D channel of the same basic access (BA) are totally separate from each other. So, a terminal on a B channel could call a terminal connected to the D channel of the same BA, although this may not happen in reality.

The layer 2 protocol for the B channel is LAPB, whereas the layer 2 protocol for the D channel is LAPD. Thus the protocol used to set up a link (in the case that on–demand link setup is provided) to the packet handler (PH) is carried over the D channel in an LAPD protocol. Only one LAPB data link at a time can be set up on the B channel, but several LAPD data links can be set up on the D channel. Data links from different terminals on the D channel are distinguished by the terminal endpoint identifier (TEI) field located in the address field of the frame (see Figure 2.19). Different types of datalinks — because besides data links for the data conversation datalinks for signaling, circuit switching, and management are also required — are distinguished by the SAPI (service access point identifier). SAPI values can be 0, 1, 16, and 63 as explained in section 2. The X.25 protocol is applied on layer 3. The ISDN network is transparent for the layer 3, so the packet-switching subscriber is in fact considered a subscriber of the PSPDN.

3.3.2.2. The ISDN Subscriber Access

The different building blocks that can form a network are depicted in Figure 3.18.

The ISDN subscriber module (ISM) is the module in the local exchange to which the BAs are directly connected. It will perform the layer 1 and layer 2 functions for B and D channels. For the B channels this consists in passing on the B channels transparently to the digital trunk module (DTM). For the D channels it performs a frame-switching function and a multiplexing of several D channels into one channel. This one channel is then called the *Bd channel*.

The ISDN remote subscriber unit (IRSU) and interface module (IRIM) together perform the same function as the ISM. A subscriber connected to the IRSU looks the same to the exchange as if it were connected to the ISM. However, because of the 2 Mb/s interface between IRSU and IRIM and because of the concentration function in the IRSU for D channels, a frame-relaying function is provided in the IRSU as well as in the IRIM. The multiplexing of D channels is done in the IRSU.

3.3.2.3. The ISDN Packet Trunk Module (IPTM)

In the local exchange (incoming side), the IPTM performs the function of PRA (primary rate access) interface to the PABX (private automatic branch exchange). As the PABX may not contain a frame-relaying function, this is provided by the IPTM. D channels are multiplexed in the IPTM and eventually multiplexed again in the IPTM at the outgoing side. B channels are passed semipermanently to the DTM. On the outgoing side of the local exchange this module performs, in addition to the function of a normal trunk module, the frame relaying and further multiplexing of D channels. Packet calls over B channels are normally passed over the DTMs, but if they are switched to the IPTM, which is possible because the IPTM also performs a normal DTM function, then the B channels are transmitted semipermanently to the next exchange and are not further multiplexed.

In the transit exchange the IPTM performs the function of packet handler interface (PHI). It talks to the packet handler over a modified PRA interface. For the B channels there is still a physical relation between one B channel at the subscriber's BA and the inlet in the PH. For the D channel this physical relation is lost because of the different multiplexing points in the network. There could be more than one transit exchange in one call, but these leave B as well as D channels unchanged and pass them semipermanently to the next exchange.

3.3.3. ISDN Packet-Mode Bearer Services (PMBS)

Two classes of ISDN packet-mode bearer services exist:

1. ISDN virtual call (VC) and permanent virtual circuit (PVC) bearer services provided on the B channel of the basic access and primary rate access
2. ISDN virtual call (VC) and permanent virtual circuit (PVC) bearer services provided on the D channel of the basic access and primary rate access

3.3.3.1. ISDN Virtual Call (VC) and Permanent Virtual Circuit (PVC) Bearer Services Provided on the B Channel of the Basic Access and Primary Rate Access

These packet mode bearer services allow users (such as terminals in a point–to–point communication configuration) to communicate via the ISDN using X.25 encoding, by means of procedures over a B channel in both directions continuously and simultaneously for the duration of a call. User classes 8 – 11, 13, and 30 as specified in CCITT Recommendation X.1 can be supported.

Provision and Withdrawal

Packet mode bearer services are provided on a subscription basis.

- *General subscription to the B channel packet service.* Some networks may not require specific subscription to packet service because it may be offered to all ISDN subscriptions. X.31 requires terminals to be identified by E.164 numbers.
- *Subscription to a service profile.* The following standard user service profile is defined to be applicable to users who are not registered in the network against any specific user's service profile applied at subscription time:

  - single link procedure, modulo 8
  - standard basic packet sequence numbering (modulo 8)
  - incoming and outgoing calls allowed
  - two two-way logical channels
  - default maximum packet length: 128 octets
  - default window size: 2
  - fast select acceptance facility
  - defaults throughput class: A (9600 bit/s)
  - throughput class negotiation facility available
  - transit delay negotiation allowed.

Normal Procedures

Activation, Deactivation, Registration. Subscription to the multiple subscriber number (MSN) service, or direct dialing in (DDI) supplementary services may be required for terminal selection purposes.

Invocation and Operation. Virtual call and permanent virtual circuit procedures can be invoked and operated concurrently by a given terminal.

First, in the case of virtual call procedures layer 1 may be permanently active but is normally activated on demand by the DTE or the packet handler (PH) via signaling. The procedures for layer 1 activation depend on methods for layer 2 activation; see the following sections. For primary rate access (PRA), layer 1 is permanently active. A B channel connection to the packet handler can be established permanently or on demand. In the latter case normal Q.931 signaling procedures are used by the terminal or the network to establish this channel.

Two different methods of layer 2 activation can be identified for the B-channel :

1. Method 1: Semipermanent layer 2.
   An X.25 LAPB link is activated at subscription time by means of an operator command (that is, there is no signaling). The network keeps the data link in the activated state. A permanent B channel connection to the packet handler is required.
2. Method 2: On-demand layer 2.
   An LAPB link is activated, in accordance with recommendation X.25, over the B channel (see Figure 3.19). Its activation is initiated by either the DTE or the network using circuit-switching principles, as illustrated in Figure 3.20.

Terminal selection and identification for the multiple subscriber number (MSN) and direct dialing in (DDI) supplementary services are available. The user is not allowed to accept a new call-offered Q.931 procedure on an already established B channel. Successive incoming calls to the same ISDN number will be directly multiplexed on an already established B channel, irrespective of information contained in the X.25 called-address extension facility field. However, in the case of no notification class, the packet handler could make use of this information to identify a specific B channel (see Annex G of Recommendation X.25). This would be a nonstandard use of a facility intended to support the OSI network service and would remain a network option.

For call establishment, Q.931 procedures are used to notify the user of incoming calls. X.25 packet layer call establishment procedures are operated on an active LAPB link. There is only one active LAPB link per B channel.

For data transfer and termination of the call, X.25 packet layer data transfer procedures and call clearing procedures apply.

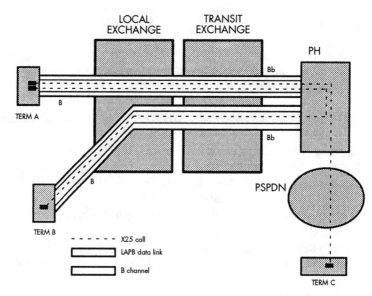

Figure 3.19 Circuit-switched connection between the terminal and the PH.

Q931		Q931	ISUP		ISUP	Q931*		Q931*
LAPD		LAPD	MTP		MTP	LAPD		LAPD
Dch		Dch	Trun		Trun	Dch		Dch

TERMINAL   LOCAL EXCHANGE   TRANSIT EXCHANGE   PH

Figure 3.20 Protocol stack (outband signaling plane).

For layer 2 deactivation, the terminal or the network should deactivate the LAPB link after clearing the last VC unless layer 2 is semipermanent or follow-on calls are expected.

For layer 1 deactivation and channel release, after clearing the last VC, the terminal or the network should release the established B channel, unless it is semipermanent or follow-on calls are expected. Layer 1 should be deactivated (from the network side only) if it is not needed by other services. However, it has to remain active for semipermanent layer 2. (Note: Primary rate access (PRA) has no defined deactivated state.)

Second, in the case of permanent virtual circuit procedures, layers 1 and 2 must be permanently active, and the B channel has to be established at subscription time. (Some networks may offer PVCs by using on-demand connections.) The terminal selection and identification is also fixed at subscription time. Call establishment and call termination are not applicable. Data transfer is executed following X.25 packet layer data transfer procedures.

3.3.3.2. ISDN Virtual Call (VC) and Permanent Virtual Circuit (PVC) Bearer Services Provided on the D Channel of the Basic Access and Primary Rate Access

These services are generally available on point–to–multipoint (for example, passive bus) and point–to–point ISDN access arrangements. These packet mode bearer services allow users (terminals) in a point–to–point communication configuration to communicate via the ISDN using X.25 encoding, by means of procedures over a D channel in both directions continuously and simultaneously for the duration of a call. User classes 8 – 10 as specified in CCITT Recommendation X.1 can be supported on the basic access (D channel at 16 kbit/s), use of class 11 may be available if the primary rate access (D channel at 64 kbit/s) is offered.

Provision and Withdrawal

These services are provided on a subscription basis:
- *General subscription to the D channel packet service.* Some networks may not require specific subscription to the packet service because it may be offered to all ISDN subscriptions. X.31 requires the terminals to be identified by means of E.164 numbers.
- *Subscription to a standard service profile.* The following standard user service profile is defined to be applicable to users who are not registered in the network against any specific user's service profile applied at subscription time. Support of the OSI Network layer service is a general requirement of this standard service profile.

  - single link procedure, modulo 128
  - standard basic packet sequence numbering (modulo 8)
  - incoming and outgoing calls allowed
  - two two–way logical channels
  - default maximum packet length: 128 octets
  - default window size: 2
  - fast select acceptance facility
  - default throughput class: A (9600 bit/s)
  - throughput class negotiation facility available
  - transit delay negotiation allowed

Normal procedures

Subscription to the multiple subscriber number (MSN) or direct dialing in (DDI) supplementary services may be required for terminal selecting purposes. Virtual call and permanent virtual circuit procedures can be invoked and operated concurrently by a given terminal.

Virtual Call Procedures.

Layer 1 may be permanently active but is normally activated on demand by the DTE or the packet handler. For primary rate access (PRA) layer 1 is permanently active. All packet information is conveyed in logical links identified by SAPI = 16. Each terminal has its own logical link (SAPI = 16) identified by TEI value. Three different methods of layer 2 activation can be identified:

1. Method 1: Semipermanent layer 2. Logical links between the DTE and the network (packet handler) are activated at subscription time by an operator command. The network keeps the data link layer in the activated state.
2. Method 2: On-demand layer 2 with fixed TEI values, also call PLL (permanent logical link). The TEI of the terminal is assigned at subscription time (known by the network); that is, manual TEI assignment is used. The activation of a logical link is initiated either by the DTE or the network depending on the direction of the first virtual call. No Q.931 procedures are used. The data link is setup by using out-of-band signaling, as shown in Figure 3.21.

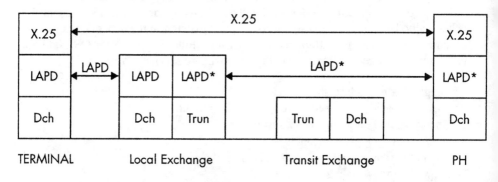

Figure 3.21 Protocol stack (data + outband signaling plane).

3. Method 3: On-demand layer 2 with dynamic TEI allocation. This method is restricted to one logical link in the case of point–to–point access arrangements.

The activation of a logical link is initiated either by the DTE or the network, depending on the direction of the first virtual call. In the case of incoming calls (network to user), Q.931 call-offering procedure (conditional notification class) may be used to interrogate the layer 2 address (TEI) to be used for the call.

Terminal Interface Identification (Network to Terminal)

This description deals only with a single user-network interface supporting multiple logical links. Users can operate several packet terminals in their in-house installation. In general, an ISDN number is used to identify a user access. In addition, the multiple subscriber number facility may be used, there by allowing users to allocate a specific ISDN number to a given terminal or terminal adapter.

Successive incoming calls to the same ISDN number will be directly multiplexed on an already established logical link, irrespective of information contained in the X.25 called-address extension facility field. However, in the case of no notification class, the packet handler could make use of this information to identify a specific logical link (see Annex G of Recommendation X.25). This would be a nonstandard use of a facility intended to support the OSI network service and would remain a network option.

In addition to these methods, additional digits from the X.121 numbering scheme can be allocated to a user, as described in CEPT Recommendation T/CD 08–03.

Terminal Interface Identification (Terminal to Network)

In the case of dynamic assignment of TEIs and use of MSN or X.121 subaddresses, the terminal identity is derived from the first call request after successful activation of layer 2. In this case the terminal must provide its identity immediately after layer 2 activation otherwise a call to the terminal may not be successful.

Call Establishment

Q.931 procedures may be used to notify the user of incoming calls. X.25 packet layer procedures are operated on an active logical link. A call is established over the whole network, as shown in Figure 3.22. In this example, terminals A and B use the same D channel and each have their own packet data link. Terminal A also has a signaling data link that the subscriber can use to establish an extra data link on demand. The packet data links are switched in the local exchange and multiplexed into 64 kb/s channels (called Bd channels) toward the packet handler (PH). An X.25 call runs between applications of terminals B and C.

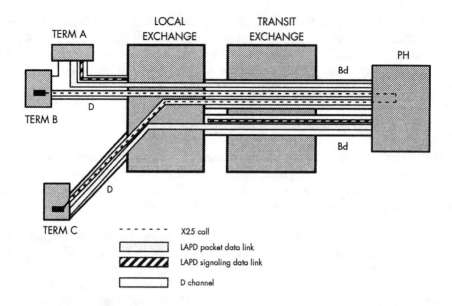

Figure 3.22 Call between D channels.

For data transfer and call termination, X.25 packet layer procedures apply. The terminal or network should deactivate layer 2 after clearing the last VC unless layer 2 is semipermanent or follow-on calls are expected. Layer 1 should be deactivated (from the network side) if it is not needed by other services. However, it has to remain active for semipermanent layer 2. (Note: Primary rate access (PRA) has no defined deactivated state.)

Permanent Virtual Circuit Procedures

Layer 1 must be permanently active, and Layer 2 must be permanently available. Terminal selection and identification are fixed at subscription time. Call establishment and termination are not applicable. Data transfer follows X.25 packet layer data transfer procedures.

3.3.4. Signaling on the Packet Handler Interface

3.3.4.1. Signaling for Bb and Bd Channel Establishment

Incoming Calls

Bb Channel Establishment

1. X.31 Case A. The PH initiates a circuit mode setup using the called party number of the subscriber.
2. X.31 Case B. The PH initiates setup of a Bb channel with the called-party number of the subscriber, and bearer capability is packet mode. Layer 2 is specified as X.25 LAPB. The PH selects the time slot in the PHI to be used for the Bb channel. The transit exchange will proceed with a Bb channel setup to the subscriber.

Bd Channel Establishment

The PH initiates setup of a Bd channel with the called-party number of the subscriber, and bearer capability is packet mode. Layer 2 is specified as LAPD. The PH selects the time slot in the PHI to be used for the Bd channel. The transit exchange will set up a Bd channel across the network with a called-party number equal to the subscriber's directory number.

Outgoing Calls

Bb Channel Establishment

1. X.31 case A (circuit mode). The ISDN subscriber specifies the E.164 number of the PH port as the called-party number, and bearer capability is circuit mode. The transit and local exchange will use circuit mode setup procedures for the Bb channel establishment.
2. X.31 case B (packet mode). The ISDN subscriber indicates a B channel access and the bearer capability is packet mode. The called-party number is not included. The transit exchange will select a trunk group to the PH and will initiate a Bb channel setup. Bearer capability is packet mode, and X.25 LAPB is layer 2. The PH decides which time slot in the PHI to use as Bb channel.

Bd Channel Establishment

1. The subscriber initiates a Bd channel establishment (see Figure 3.22). The subscriber requests D channel service, and a new Bd needs to be set up. The local exchange will

initiate a Bd channel setup to the transit exchange using the E.164 number assigned to the PH. It is proposed that "Transmission Medium Requirement" and "User Service Information" parameters be used in SS7 to indicate 64 kbps unrestricted for Bd channel establishment. This E.164 number is unique for the transit exchange to PH association. Once the call terminates on this number, the transit exchange will initiate Bd channel establishment over the PHI. Once the Bd channel is established, the local exchange will initiate the Bd channel management information exchange.
2. The network initiates a bd channel establishment (Figure 3.22). The local exchange can initiate a Bd channel establishment independent of any subscriber request if no Bd channel is available and an in-band signaling data link is required. The same procedures apply as when the subscriber initiates the channel establishment except the subscriber initiation.

3.3.4.2. Signaling for Bd Channel Parameter Exchange

Establishment of Signaling Link

The signaling link per bundle of Bd channels is always established from the PH direction toward the frame handler.

Bd Channel Parameter Exchange for Incoming Calls

For an incoming D channel call with established Bd channels, the PH sends a setup message to request DLCI (data link connection identifier) assignment and Bd channel reference. If the FH provides frame relay function, the call reference is retained for a subsequent release of layer 2.

Bd Channel Parameter Exchange for Outgoing Calls

For an outgoing D channel call with established Bd channels, the FH sends a setup message to PH indicating DLCI and Bd channel reference.

3.4. PACKET-SWITCHING NETWORK ASPECTS

3.4.1. Why Bother with Packet-Switching?

Why is circuit switching alone not enough to transport data in the ISDN network? Indeed, the network would be simpler if we had only one signaling mechanism, one rate structure, one numbering plan, a 64 kb/s transparent pipe, and one network. Moreover, all the principles that can be used are very similar to the ones used for telephony. However, for many data services such as point of sale and credit card checking, the 64 kb/s bandwidth of a B channel is a waste of resources. Many of the existing terminals need only 1.2 to 2.4 kb/s for normal operation. In some European countries more than 80 percent of the traffic in the data network uses speeds lower than 2.4 kb/s.

Furthermore, the needed quality of transmission may not always be available. Presently three information transfer capabilities exist in circuit mode: (1) voice quality, (2) modem quality, and (3) bit integrity quality. For the first two types the quality may be insufficient for the data unless sophisticated subscriber equipment is used. The third type is available only in a network that is fully digital from terminal to terminal.

If packet-switching is used, bit integrity quality is guaranteed. Furthermore, the network can be used in a much more efficient way because the D channel can also be used for data transfer and because calls can be multiplexed on one link so that the silence periods on the channels is decreased.

3.4.2. Why Maximum Integration?

For the ISDN subscriber

- Only one subscription is required with only one number. In the case of minimum integration it is more difficult to structure the numbering plan because X.121 and E.164 numbers have to be mixed. With maximum integration, a single numbering system encompassing voice and data traffic can be worked out, and number conversion is not required.
- No split charging is required. In the case of minimum integration, the charging has to be split between ISDN and PSPDN. Intrinsically the charging for packet-switching services in fully integrated ISDN could be a small delta on top of the voice charges, which could be a requirement to bring ISDN in the premises of the nonbusiness user.
- It becomes possible to line up call facilities between data (circuit and packet) and voice services. Thus, for instance, subscriber-control operations for the activation or deactivation of call diversion can have the same command format for both applications.

- Some call-offering options (as specified in CCITT recommendation X.31) are possible only with an E.164 numbering scheme.

For the network operator

- To calculate the extra bandwidth needed for packet-switching, suppose that all ISDN subscribers use packet-switching and take into account the multiplexing possibilities of packet-switching. One arrives at a figure of roughly 4 percent. For this increase in bandwidth, no extra equipment has to be installed except for the modules containing the software for packet or frame handling.
- Fully integrated packet-switching allows a higher flexibility of extensions, because when extending an exchange with telephone lines at the same time, enough capacity can be added for treating the packet switching. In the case of minimum integration, the packet handler has to be extended every time when an extension is done anywhere in the network. This means not only that extensions need to be done at two places but also that the PH has to be modified very often, which creates the risk of instability in performance.
- Only one operation and maintenance center has to be used (instead of two), and coordination of given tasks between two OMCs is usually not an easy job.

3.4.3. Disadvantages of a Minimum Integration Network

The packet handler interface (PHI) as specified by ETSI has the following disadvantages (see Figure 3.23):

- A duplication of functions is present between D channel signaling used to set up circuit-switched connections to the PH and the X.25 in–band call setup protocol.
- Double subscriptions are required. An ISDN subscriber with packet-switching services must subscribe to the ISDN and PSPDN. The PSPDN needs to handle E.164 numbers.
- Double-charging mechanisms have to be applied, and a principle of division of revenue has to be selected, not only for calls but also for supplementary services.
- An available service has to be supported over two networks, each with its own rules. For example, a closed user group created in the ISDN for a group of circuit-switched subscribers may work differently than when the same subscriber belongs to a closed user group that is supervised for packet-switching subscribers by the PSPDN.
- Call offering procedures, widely used and available in the telephony networks and now also in ISDN, are not lined up with the X.25 throughput negotiations that define the data rate of the data connection.
- Because the PHs are not available in positions as close as possible to the subscribers, a star network will have to be built, and fully built-out meshed networks are not possible. (see Figure 3.24).

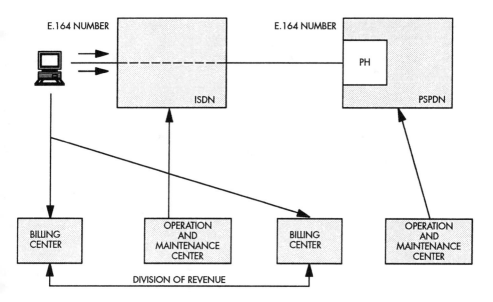

Figure 3.23 Overlapping functionality of two networks.

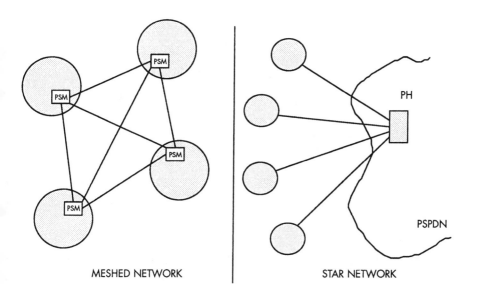

Figure 3.24 Meshed networks and star networks.

- The reliability and performance figures stating the requirements on a per-call basis still have to be split over the two networks. In general, however, an overall lower quality will result if the reliability and performance of each of the networks is not improved in comparison to the one network case.

One practical but sometimes blocking consideration in the selection of minimum or maximum integration is the mere availability of a PSPDN. In many countries a data network of any sort is in fact not present. This is so not only in less developed countries but also in countries that have a large geographical spread of cities. And if a PSPDN is available, the addition of PHs to serve the whole country will create a considerable overhead (see Figure 3.25).

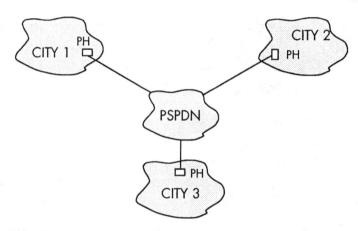

Figure 3.25 PH distribution in the network.

3.4.4. Material Cost Difference

The frame handling required for the treatment of D channels in the minimum integration case already provides the necessary hardware to perform packet-switching. (At least this is so in systems with a distributed control architecture). It is necessary only to add more firmware and software (see Figure 3.26). Thus, the material is intrinsically available but not used.

In the minimum integration case the PH performs the packet-switching function, thus, this hardware is the cost difference between minimum and maximum integration.
COST MIN = FRAME HANDLERS + PH
COST MAX = PACKET HANDLERS (the packet-switching modules in the ISDN)
FRAME HANDLERS = PACKET HANDLERS

Thus the cost difference is PH. (All the equipment identical in both cases is left out of the equation.)

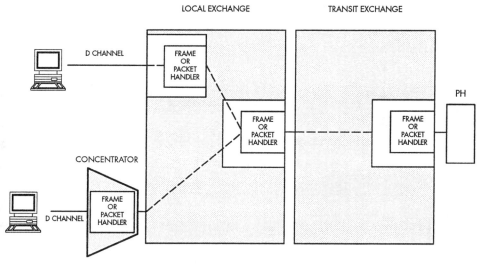

Figure 3.26 Material cost.

Not only is the minimum integration network more expensive, it is also less efficient in using its resources. A simple example will illustrate this easily. If two subscribers connected to the same local exchange —say, two centrex subscribers located in the same group— want to establish a packet-switching connection, then they have to setup a call to the PH and back again (see Figure 3.27). In the maximum integration case on the other hand, the call remains in the local exchange.

3.4.5. Extra Functions Needed to Convert a Minimum Integration Network into a Maximum Integration Network

When calls have to be routed to the PSPDN from an ISDN network that performs its own packet-switching, then analysis of X.121 has to be possible. X.25 call control must be performed in ISDN, and the X.75 has to be installed between ISDN exchanges and between the ISDN network and the PSPDN. The layer 2 functionality has to be extended, because in the intermediate steps in the network the frame treatment has to be enhanced. For instance, Q bit and D bit handling and segmentation of packet and reassembly of packets need to be done.

Lining up of circuit- and packet-switching supplementary services can be achieved for closed user groups, call forwarding, call deflection, calling line identification presentation, and calling line identification restriction. All these extra functions would not need a higher effort to implement them than the one needed for implementing frame handlers and a packet handler interface in the minimum integration case.

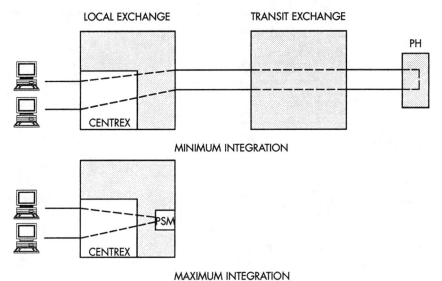

Figure 3.27 Packet-switching between two centrex subscribers.

3.4.6. Quality of Service Comparison

Data transfer delays for minimum and maximum integration are identical: plus or minus 25 msec. for an average packet length of 128 bytes. The call setup delays for the first packet-switched calls are identical for minimum and maximum integration. For subsequent calls the following differences can be noted:

- lower setup delays for an outgoing with minimum integration
- equal call setup delay for incoming calls

In a connection between two networks (PSPDN and ISDN), a higher failure rate can be expected because more HW elements are used in the connection than if the call stays in one network only. Division of revenue techniques have to be applied to split the taxation between two networks (PSPDN and ISDN) with a probable higher loss of charging information, because the charging data has to be treated by many more network elements than if only ISDN is involved. Fault localization over two networks will last longer, with a possible consequence of a higher lost call rate than if only one network has to be diagnosed.

3.4.7. Disavantages of X.25

X.25 has the merit to exist and to be widespread in several countries in the world. See Appendix A.2 for a more detailed discussion. However, in comparison to other upcoming techniques, the following disadvantages should be noted.

Disadvantages of X.25 in General

- The X.25 is a well-established protocol, but this also means that it is becoming out of date, and over the years several variants have arisen. The same is true for X.75, and therefore the latest requirements from networks that do evolve cannot be completely fulfilled.
- The X.25 protocol has an unnecessary overlapping between layers 2 and 3, which makes it less efficient than it could be.

Disadvantages of X.25 for ISDN Subscribers

- The subscriber terminal needs an additional board and a complementary SW package to execute the layer 3 protocol. Because many X.25 protocol variants exist the available products on the market have to compromise on the supplied functionality.
- The subscription profiles that are required for X.25 data calls are not all supported by the ISDN network, because the X.25 is in fact a matter between the subscriber and the PH. Therefore the applications (or software running in the subscribers equipment) has to know the profiles of the called subscriber in order to select the right match of profiles between A and B subscribers.

3.4.8. Interexchange Signaling Aspects (see Figure 3.28)

When packet-switching is provided in ISDN, which is the best signaling system between exchanges of the ISDN network and between different data networks? X.75 is a candidate and is used today, but it has several disadvantages. X.75 is an internetwork protocol and is not optimal as an interexchange protocol. It is incomplete in the support of extra subscriber facilities and charging options. Furthermore, the network management aspects are weak: not only do diagnostic and alarm information have to be transferred to the supervision point, but the recovery for established calls also has to be supported (and that is not the case). On the other hand, X.75 remains required for international data call setup as long as no other signaling system is adapted for it.

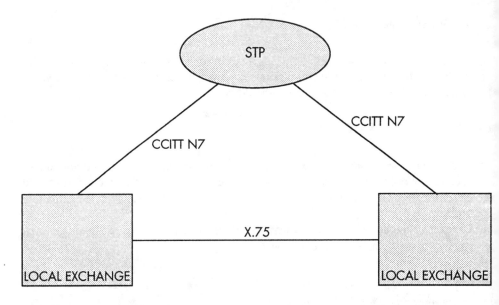

Figure 3.28 Interexchange signaling aspects.

The CCITT N7 system is an interexchange, intranetwork protocol that also offers several internetwork options such as ITUP and the international ISUP. It also offers more inherent features applicable for good network management, such as TCAP, SCCP, and OMAP. Because the CCITT N7 is the basic signaling in ISDN for voice and circuit-switching calls, little extra functionality is needed to support packet-switching. CCITT N7 will also very likely be the basis for the interexchange signaling system for future broadband networks. This broadband network is basically a packet-switched network. Therefore any change in signaling systems should move in the direction of the common channel signaling and out-of-band system CCITT N7.

3.4.9. Conclusion

From the above the following conclusions can be drawn:

- Packet-switching is required.
- It should be made available completely integrated within the ISDN.
- It can be introduced without extra costs to the network.
- It should not necessarily be based on X.25 and X.75.

To better integrate packet-switching into ISDN, it is normal to line up with the available functions of voice and circuit switching which are as follows:

- out-of-band signaling for call setup
- in-band support of supplementary services, which should line up between the different bearer capabilities

With transmission systems becoming more and more reliable the data transfer functions of layer 3 can be left to the subscriber equipment.

The above concepts are the basis for the standardization of a new packet mode (NPM), also called additional packet mode bearer service (APMBS). This NPM will be based on frame relaying. Frame relaying, like ATM, is becoming the name of the product. Indeed, frame relaying is not the only technique used; in parts of the network frame-switching principles are also employed. Furthermore, besides the layer 2 functions defined by "frame relaying" for data transfer during the data conversation, a protocol is needed for call setup supported between exchanges by a modified CCITT N7 signaling. Not only this NPM will support 64 kb/s speeds, but those systems that can allow it will also switch 2 Mb/s channels.

Besides the NPM, which is not yet fully standardized, another possible packet-switching technique that has much in common with the new packet mode in terms of simplicity of principles, integration into ISDN, and the fact that it is based on CCITT N7 is the User Signaling Service 4. It is, however, not fully comparable with the NPM because it is meant more as a complement of the X.25 services at the low-speed end. The offered throughput is around 1 kb/s. This protocol could easily cover applications such telemetry, fax, message services, and the like.

3.5. USER-TO-USER SIGNALING

3.5.1. Packetized Signaling in ISDN Networks

The natural emphasis in this book is on techniques for support of X.25 packet-switching in the ISDN, with discussion of evolutionary aspects such as NPM or the future broadband ATM. The book would not be complete, however, without some mention of what has been called the largest packet-switching network in the world. This has nothing to do with X.25 and is called CCITT Signaling System Number 7. It switches packetized messages between exchanges and other specialized centers in the PSTN-ISDN-PLMN and may be considered the nervous system of these networks.

We should also mention the use of packetized data messages for user–network signaling on the ISDN access as described in CCITT Digital Signaling System Number 1. When these messages carry data transparently between subscribers, then we have essentially another packet mode bearer or supplementary service offered by the ISDN. In this chapter we discuss transfer of signaling messages between the user and the network, between network nodes, and between users themselves. In a general sense, these are all packet-switching mechanisms fully integrated into the ISDN.

3.5.2. Architecture

User–network signaling, user–to–user signaling, and common channel signaling capabilities are discussed here. Figure 3.29 shows the basic architectural model of an ISDN, taken from CCITT Recommendation I.325. Although Figure 3.29 makes a distinction between packet–switching and signaling capabilities, the following considerations justify generalization for the purposes of this chapter:

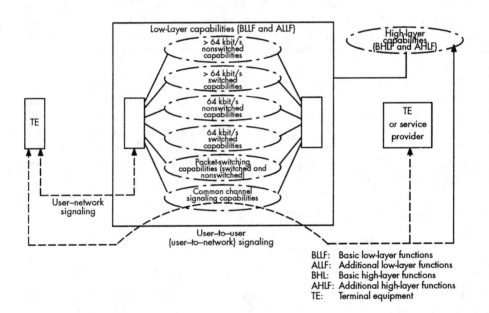

Figure 3.29 Basic architecture model of an ISDN.

- On the user–network interface, the transfer of frames containing signaling messages between the exchange and identified terminal equipment can be regarded as a form of local packet-switching. In fact, extension of this layer 2 routing capability forms the basis for the NPM.
- CCITT Recommendations I.230 and I.232.3 describe user–to–user signaling unrelated to a circuit as a packet-mode bearer service category, distinct from the virtual call and permanent virtual circuit categories for X.25 support in the ISDN. When user–to–user

signaling is associated with a circuit, however, it is regarded by CCITT Recommendation I.257 as an additional information transfer supplementary service.
* Between exchanges and other centers, CCITT Recommendation Q.700 describes Signaling System Number 7 as a form of data communication that is specialized for various types of signaling and information transfer within telecommunication networks. Data communication is achieved by packetized messages.

All of these involve the transfer of labeled packetized data between identified entities in the ISDN, so that packetized signaling entails a form of packet-switching in the ISDN.

3.5.2.1. User–Network Signaling

LAYER 2

On the ISDN user–network interface, the D channel is used for both signaling and other purposes. HDLC framing is used, and the LAPD frames have the address field header shown in Figure 3.30, taken from CCITT Recommendation Q.921. This scheme allows for multiplexing of various datalinks on the single D channel, with the TEI used to identify the user terminal and the SAPI used to identify the function within the terminal. Each datalink connects the terminal function with a peer function, between which LAPD procedures operate. Transfer of frames between these peers is a frame relay mechanism and therefore, in general, a kind of packet-switching. In fact, when the peer function resides in another user terminal and a single terminal uses multiple TEIs, we have the full–blown layer 2 packet-switching (frame relay) of the NPM. Such frames are identified by SAPIs defined in CCITT recommendation Q.922.

8	7	6	5	4	3	2	1	
SAPI						C/R	EA 0	Octet 2
TEI							EA 1	3

EA: Address field extension bit
C/R: Command-response field bit
SAPI: Service access point identifier
TEI: Terminal endpoint identifier

Figure 3.30 LAPD address field format.

For other SAPIs, the layer 2 switching is more limited, involving concentration and discrimination. For SAPI = 16, the peer function is a layer 2 entity used by an X.25 layer 3 packet handler. Calls are multiplexed onto each datalink using X.25 packet layer procedures, each call being identified by a logical channel in packets conveyed as layer 2 user data. For SAPI = 0, the peer function is a layer 2 entity used by Q.931 call control. Calls are multiplexed onto each datalink using Q.931 layer 3 procedures, each call being identified by a call reference in signaling messages conveyed as layer 2 user data. We will now discuss these messages further.

LAYER 3

The Q.931 signaling messages have the general layout shown in Figure 3.31, taken from CCITT Recommendation Q.931. The protocol discriminator normally has a fixed value, but in certain countries it has been used to distinguish versions of Q.931. The call reference is used to identify the call and allow multiplexing of several simultaneous calls onto the same datalink. The message type identifies the message. For our purposes, the most important messages are as follows:

SETUP	to initiate call establishment
ALERT	to indicate that the called user has been alerted
CONNECT	to complete call establishment
DISCONNECT	to clear an access circuit connection
RELEASE	to initiate call release
RELEASE COMPLETE	to complete call release
USER INFORMATION	to transfer information between users
CONGESTION CONTROL	to control flow of information between users

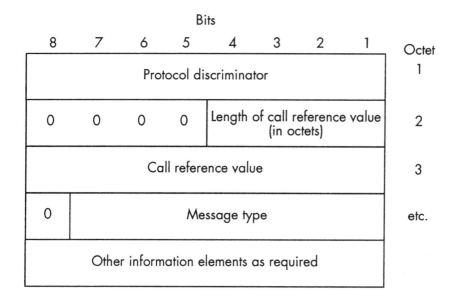

Figure 3.31 General message organization example.

Other information elements are organized for the most part as tag–length–value objects. For our purposes, the most important information elements are as follows:

bearer capability	to request a particular bearer service
cause	to specify cause for clearing or other events
channel ID	to identify channel(s) used for the call
called party number	to identify the called user
user–user	to convey transparent information between users
congestion level	to indicate receiver ready or not ready

A simplified view of Q.931 procedures suffices here. In what follows the forward direction is from calling to called user (establishment) and from releasing to released user (release).

1. Call establishment

    SETUP (called party number, bearer capability, channel id, user–user) is sent forward to initiate call establishment.

    ALERT (user–user) may be sent backward to indicate called user alerting (ringing for example).

    CONNECT (user–user) is sent backward to complete call establishment.

2. Active phase
   If B channel(s) have been established, they can now be used for communication between the users.
3. Call clearing
   If B channel(s) have been established, a three-message clear is performed:

   > DISCONNECT (cause, user–user) is sent forward.
   > RELEASE is sent locally backward.
   > RELEASE COMPLETE is sent locally forward.

   If no B channel has been established, a two-message clear is performed:

   > RELEASE (cause, user–user) is sent forward.
   > RELEASE COMPLETE is sent locally backward.

This simplified scenario is shown in Figure 3.38, which also shows resulting messages between exchanges.

3.5.2.2. User–to–User Signaling

The preceding simplification of Q.931 procedures shows transfer of data between users piggybacked in signaling messages. In addition, data may be carried between users in specialized USER INFORMATION messages. This amounts to a packet–switching capability fully integrated in the ISDN. The capability may or may not be associated to a circuit–switched connection involving B channels. This distinction is functionally rather unimportant from the point of view of the user. However, the distinction is important in terms of the ISDN, because the service is regarded as supplementary in the first case (being associated to a bearer service) and as a bearer service in the second case (not being associated to another bearer).

User–to–User Signaling Associated with Circuit–Switched Calls (see Figure 3.32)

The user data transfer capability is decomposed into the following user–to–user signaling (UUS) supplementary services:

- Service 1 (UUS1): transfer in signaling messages exchanged to establish or release the circuit call
- Service 2 (UUS2): transfer in USER INFORMATION messages between ALERT and CONNECT
- Service 3 (UUS3): transfer in USER INFORMATION messages in the active phase (ie after CONNECT)

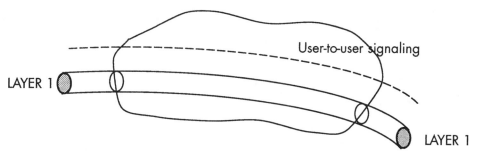

Figure 3.32 User–to–user signaling.

For UUS2, up to two USER INFORMATION messages in each direction are allowed. For UUS3, any number of USER INFORMATION messages in each direction are allowed. However, the rate of submission of these messages is limited and may be further restricted (for example, in the case of network congestion) by use of the CONGESTION CONTROL message.

User–to–User Signaling not Associated with Circuit–Switched Calls

Because there is no associated circuit, this counts as a bearer service, identified by CCITT Recommendations I.230 and I.232.2, as the "user signaling packet-mode bearer service category," or User Signaling Bearer Service (USBS) for short. The SETUP for a USBS call (called a temporary signaling connection in Q.931) is distinguished by a specific value of bearer capability and channel ID. DISCONNECT is not used because there is no access circuit connection (B channel) to release. Length restrictions on the user–user information element in the USER INFORMATION message are relaxed (compare to connectionless packet-switching). Apart from these exceptions, the service is the functional equivalent of the sum of UUS1 and UUS3 without a circuit. For the main part we ignore the distinction from now on and consider the service characteristics and possible applications of user–to–user signaling without considering whether or not a circuit happens to be associated.

3.5.2.3. Comparison to X.25

Table 3.1 compares user–to–user signaling with X.25 support in the ISDN, from the point of view of the user.

FEATURE	USER–TO–USER	ISDN X.25
Procedure definitions	Q.931	Q.931 X.31 X.25
Supplementary services	ISDN	ISDN X.25 facilities
Subscription and classes	ISDN	ISDN PSPDN
Channels used	D only	D or B
Throughput classes	1 only (low)	many (low-high)
Maximum packet/message sizes	1 only	2 on D many on B
Throughput class negotiation	No	Yes
Flow control par. negotiation	No	Yes
User data in signals	Yes	Yes
Call identification	Call reference	Logical channel
Reset procedures	No	Yes

Data transfer:
   Sequence numbers
   Windows
   Delivery confirmation
   Interrupt (expedited data)
   Packet retransmission
	
	No
	No
	No
	No
	No
		
		Yes
		Yes
		Yes if used
		Yes if used
		Yes if used

Table 3.1 Comparison of User–to–User Signaling with X.25 Support in the ISDN.

Thus user–to–user signaling represents a rather rudimentary packet-switched service when compared with X.25 support in the ISDN. In contrast to NPM, it is not therefore a direct competitor with the X.25 service. It is certainly not suited to bulk data transfer applications, such as file or screen transfer, which are better handled by 64 kbps circuits or high-speed X.25, NPM, or the coming ATM. It is, however, suited to transactions characterized by intermittent message transfers with a low average throughput of typically less than 1 kbps. This may sound like a restrictive class, but it does include, for example, all telephonic signaling, other interactive transactions, and many of the applications characterized as OSI remote operations.

User-to-user signaling exhibits certain advantages for support of such applications in the ISDN, when compared with other mechanisms:

- Pure Q.931 procedures are used, supportable by any ISDN terminal, without requiring extras such as terminal adaptors and X.31 or X.25 logic.
- Both circuit-related and non-circuit-related messages may be transferred.
- Sequence numbering, window rotation, reset, and negotiations appropriate to file transfers but inappropriate to intermittent message transfers are not used.
- The service is fully integrated into the ISDN, with advantages discussed in more detail elsewhere in this book: one subscription, one bill, one responsible operator, one set of supplementary services to choose from.

In addition, as we shall see, the Signaling System Number 7 network is designed to support exactly this type of intermittent message traffic between exchanges and specialized centers in an ISDN.

3.5.3. User-to-User Signaling Applications

We now discuss potential applications of the simple low-speed message transfer capability of user–to–user signaling, which may be regarded as a useful complement to X.25 and other powerful packet-switching services. First, note that all users of the ISDN are reachable by means of user–to–user signaling, including those who:

- do not wish to subscribe to any network other than ISDN
- are not prepared to buy X.25 or X.31 functionality
- do not wish to access X.25 services and subscribers
- are satisfied with a single throughput class, single message lengths, no windows, no delivery confirmation, no reset, no expedited data (interrupt) and so on
- are happy with call setup delays appropriate to telephony and message transfer delays of order 1 second

Such users would be lost to the X.25 network, and their potential data traffic lost to the ISDN.

Now refer to the basic architecture of the ISDN shown in Figure 3.29, where user–to–user signaling operates both between user terminals and between user terminals and service providers that may be embedded in the ISDN. For communication directly between users, the following applications suggest themselves:

- a cheap telex–like short-message service
- messages between smart phones, either circuit-related or non–circuit-related

- messages between ISPBXs, either circuit-related or non–circuit-related, to support the rich variety of private network facilities and other aspects of private networking
- simple access to expert systems (for example, for booking or reservations)
- messages to emergency services
- possibly, teleaction and telemetry-like services, driven from a smart phone
- point–of–sale reporting and other low–speed data applications
- circuit–related outband control of, say, file transfers on one or more 64 kbps circuit–switched connections, with functionality on the high–speed channels confined to a LAP and direct disc access

For communication between users and servers in the ISDN itself, the following applications suggest themselves:

- transfer of short messages to a server for paging or delivery to mobile subscribers
- general queries, interrogation or registration of supplementary services offered by the ISDN, in a user–friendly manner
- messages between ISPBXs and ISDN centrex groups (ISCTXs), either circuit-related or non–circuit-related, allowing creation of hybrid private networks
- messages between attendant positions and ISCTXs
- direct user access to an Intelligent Network server, capable of modifying and displaying service data, and commanding actions on the local exchange
- automated directory inquiries and other services presently requiring an operator

In fact, many of these applications are the automated equivalent of calls to the operator and would similarly entail calls to special service numbers, with interactive text messages replacing speech. Such services would be difficult to offer if they demanded X.31 or X.25 capability in addition to DSS1 capability at the user premises and in the ISDN. User–to–user signaling allows them to be offered on a global basis to ISDN users. In addition, transfer of such messages through the Signaling System Number 7 network is both feasible and natural, because ISDN (both fixed and mobile), ISCTX, and Intelligent Network applications are already users of this data communication network.

3.5.4. Signaling System Number 7

3.5.4.1. Out-of-Band Signaling

In telephony there has been a gradual movement away from in–band signaling embedded in a speech circuit or channel and toward the use of a "common" circuit or channel dedicated to transferring signaling information for separate groups of circuits or channels.

This separation is based on sound commercial and technical reasons. The use of the D channel as a common signaling (and low-speed data) channel on the digital subscriber access of ISDN is an example of this approach, but it was preceded by the use of common channel signaling for controlling trunk circuits within the PSTN. An early example of this is Signaling System Number 6, popular in the United States. The rather limited capabilities of this system led to definition of the much more powerful Signaling System Number 7, in the CCITT Q.7xx series of recommendations.

Definition of Signaling System Number 7 began at roughly the same time as CCITT Recommendations X.25 and X.75. Like them, it has a layered structure, with level 1 the physical layer, level 2 the link layer, and level 3 the signaling layer at which switching (routing) is performed. However, whereas X.25 specifies an interface between a user and a dedicated PSPDN, and X.75 specifies an interface between separate PSPDNs, Signaling System Number 7 specifies the interface between data network nodes, the nature of the network, and also defines users (at level 4) that make use of the data network to transfer messages between exchanges and other centers for signaling and other purposes. Such an intranetwork specification is absent from the X series of CCITT recommendations, so that PSPDNs internally use proprietary protocols between their network nodes, and only conform to recognized standards on their boundaries. These proprietary protocols vary between internal virtual circuit approaches (often using a freely adapted X.75–style protocol between nodes) and internal datagram approaches, in which packets are independently routed based on a destination network node identifier in their header and only associated with external virtual circuits in the boundary nodes. In OSI terms, the first approach would be termed *connection oriented*, and the second *connectionless*. For a PSPDN both have advantages and disadvantages.

The intranetwork specification of Signaling System Number 7 is called the message transfer part (MTP). It defines a connectionless but sequenced mechanism for secure transfer of packetized data between nodes, and its general services are used by a range of functions, again called "Parts." The most important of these are shown in Figure 3.33, taken from CCITT Recommendation Q.700, which provides an excellent introduction to Signaling System Number 7 for interested readers. The TUP is the first user part, and it defines pre–ISDN telephone signaling functions and messages. Note that the topology of the circuits between exchanges controlled by the TUP may be completely distinct from the topology of the MTP data communication network used to transfer the TUP messages. This gives great freedom in network design. The ISDN–UP, commonly called the ISUP, defines ISDN signaling functions and messages. When the ISUP controls circuits, the remark about topological independence for the TUP also applies. The SCCP enhances MTP capabilities by providing an OSI network layer service and by supporting the transfer of messages between any two exchanges in the world. In addition to this generalized connectionless service, the SCCP also offers connection–oriented services, again between any two exchanges. Transaction capabilities use the general connectionless service of the

SCCP to exchange operations and replies via a dialogue, supporting such applications as mobile networking and intelligent networking.

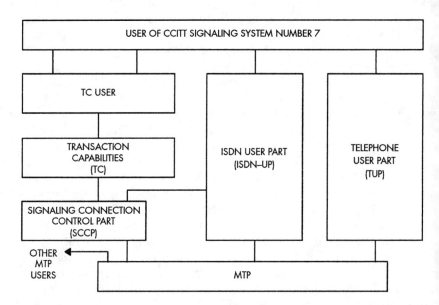

Figure 3.33 Architecture of CCITT Signaling System Number 7.

3.5.4.2. Message Transfer Part

Figure 3.34, taken from CCITT Recommendation Q.700, shows the functional levels of the MTP.

Level 1 corresponds to the physical layer of the OSI model. A 64 kbps digital path is normally used, although analogue links with modems are also possible.

Level 2 is similar to the data link layer of the OSI model. Flags, bit stuffing, and a CRC check sequence are used. The level 2 functions are essentially those of LAPB, providing for reliable transfer of level 3 data on a point–to–point link, with error detection by CRC, error correction by retransmission, congestion control, and so on. However, both the contents of frames, called signaling units, and the procedures used to achieve level 2 functions are very different from LAPB. To a certain extent, this reflects the "trunk" nature of signaling links, but is also an unfortunate consequence of parallel definition within different parts of CCITT at roughly the same time.

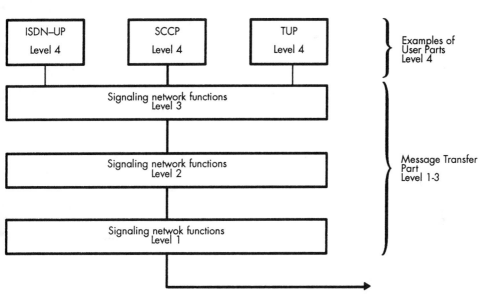

Figure 3.34 CCITT Signaling System Number 7 functional levels.

Level 3 corresponds to a connectionless network sublayer of the OSI model. It defines transfer of labeled packets, called *messages,* between users in network nodes, called *signaling points,* and also defines network management functions to control the routing and to adapt to changes in status of the signaling network and its links. A signaling point at which a message is received on one link and transferred to another link is called a signaling transfer point (STP). An STP has all the characteristics of a packet switch node.

Some networks have STPs physically separate from the exchanges and other centers that they serve: in this case, the MTP looks very much like a dedicated packet-switching network, with users distinct from the data network. Other networks embed STP functionality in the exchanges themselves, in which case the distinction between the data network and its users is logical rather than physical. This requires that the exchange architecture is integrated, that is, capable of both circuit and packet-switching. Finally, it is possible to design a network in which communicating exchanges are directly connected by signaling links, so that the STP function vanishes entirely. In this case, reminiscent of DSS1, the MTP really performs no packet-switching, and its flexibility is not used. This "associated" approach means that at least two signaling links must be equipped for each trunk circuit group and may lead to inefficiencies, both in signaling link utilization, and in transfer of non–circuit-related messages.

Figure 3.35 shows MTP switching functionality. Note that if the level 4 served by the MTP is performing circuit-switched signaling (TUP or ISUP), then exchange A is connected to exchange B by layer 1 trunk circuits (not shown). Figure 3.36 shows the general formats used at levels 2 and 3 of the MTP.

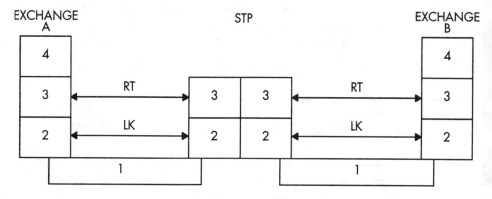

RT: Routing (+ network management)
LK: Link (delimitation, error detection, error correction, congestion control)
STP: signaling transfer point

Figure 3.35 MTP switching.

Level 2 (see CCITT Recommendation Q.703)

Each level 2 signaling unit begins with a flag, sequence numbers, and indicator bits, and (redundant) length indicator, and terminates with a CRC check sequence on the signaling unit contents. The length is used to distinguish between signaling unit types: fill-in signal units (FISUs) are idled when there is nothing else to send; link status signal unit (LSSUs) are used to establish the link (called alignment) and to indicate failure conditions and congestion; message signal units (MSUs) contain level 3 messages and are identified by sequence numbers (the forward sequence number is incremented on transmission of a new MSU, and the backward sequence number is incremented on correct reception of a new in-sequence MSU).

Two methods for error correction may be employed, namely basic and preventive cyclic retransmission. In the basic method, the backward indicator bit is changed when loss of an MSU is detected (for example, by a FISU with a changed forward sequence number); this triggers the sender to retransmit, changing the forward indicator bit. In the preventive cyclic retransmission method, typically used on satellite links, the sender cyclically retransmits unacknowledged MSUs, and the receiver simply waits for in-sequence MSUs. The idling of FISUs gives rise to a large number of interrupts unless nonstandard HDLC hardware is used. However, it does have the advantage of giving a very rapid loss detection and correction mechanism when compared with LAPB.

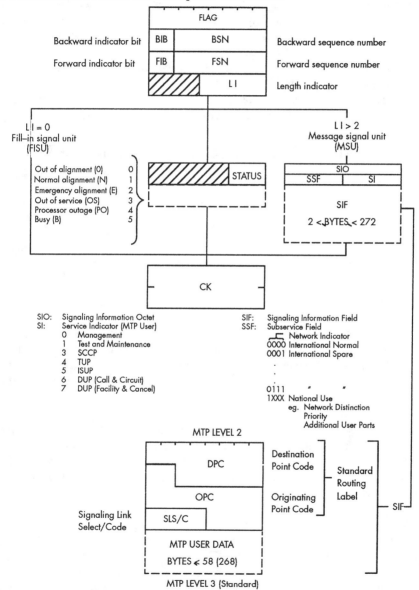

Figure 3.36 Formats for the message transfer part (MTP).

Level 3 (see CCITT Recommendation Q.704)

Level 3 of the MTP receives messages from a level 4 user in an originating signaling point, routes them to the destination signaling point, and distributes them there to the destination level 4 user. The destination exchange is specified in the service indicator of the signaling information octet that heads all messages.

Routing involves a choice of outgoing link by accessing tables at each point with the general structure shown in Table 3.2. The access of a routing table using the destination point code as key to a group of transmission links is a common mechanism in any connectionless (datagram) network. However two features of the MTP routing are worth further mention.

NETWORK	DESTINATION POINT CODE	SIGNALING LINK SELECT														
		1	2	3	4	5	6	7	8	9	A	B	C	D	E	F
		16 transmission links															
		16 transmission links															

Table 3.2 MTP Routing.

The network described above is contained in the subservice field (SSF) of the service information octet. This allows the same signaling link to carry message traffic for more than one MTP network, for instance, European international traffic and intercontinental traffic. In this case, the destination point code has meaning only when qualified with the SSF. These traffic flows can be demultiplexed by the routing tables onto separate links. In theory, this allows for up to 16 logically separate MTP networks to be overlaid, sometimes sharing physical resources and sometimes separating. In practice, because of the limits on routing table sizes, the number of networks sharing the same link is normally limited to four or less. This flexibility can also be used, for example, to multiplex internal signaling with user–to–user signaling over low utilization links, to split the traffic at higher points in the network hierarchy.

The other interesting feature is the use of the signaling link select (SLS) contained in the standard routing label to choose a specific transmission link. This means that if a user includes the same SLS in every message for a particular transaction, then these messages will arrive in sequence. For example, the TUP uses an SLS that is part of its circuit identification code, guaranteeing that all signaling messages for a given channel arrive in sequence at the destination. This sequence guarantee by the MTP means that great care must be taken when the routing tables change, for example, due to a failure. In addition,

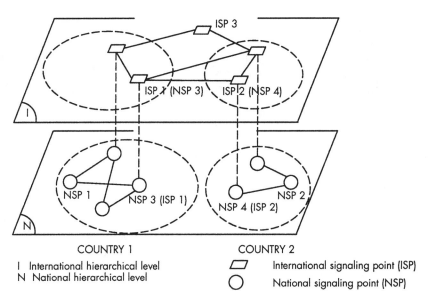

Figure 3.37 International and national signaling networks.

great care must also be taken not to lose or duplicate messages in these situations. These features make the MTP network management more complex than in a traditional datagram network in which resequencing, duplicate detection, and error correction are performed at the edges of the network. However they do simplify processing at the edges. This type of fixed routing is suited to a large number of low throughput transactions, such as the signaling that the MTP is designed to support, or the user–to–user signaling described earlier. Finally, it might be thought that the MTP routing limits to 16 the total number of outgoing links from a signaling point to a given destination point. This is indeed the case if the same routing table is used for all sources (users and incoming links) on the signaling point. However, the limit can simply be removed by making the table source–dependent.

MTP Network Structure (see Recommendation Q.705)

Within an MTP network, a signaling point (exchange or specialized center) is assigned a unique 14–bit point code, allowing for a theoretical maximum of 16,384 nodes. The allocation scheme of international signaling point codes is defined in CCITT Recommendation Q.708. Note that a node may function as both an international signaling point and a national signaling point, with distinct point codes, as shown in Figure 3.37, taken from Recommendation Q.705. The nodes shown as ISP1-NSP3 and ISP2-NSP4 exhibit this property, whereas ISP3 is pure international, and NSP1 and NSP2 are pure national. Note that the link marked X in country 1 may carry both national and international traffic, distinguished by SSF, as discussed in the preceding section.

A similar technique may be used within a country to create a hierarchy of networks, for organizational reasons, or to limit the size of routing tables. However, this obviously reduces the capability to transfer messages directly between exchanges through the MTP network. For circuit–related signaling messages, only exchanges with trunk connections require MTP connectivity. However, for non-circuit-related messages, the tendency in many countries is to configure larger MTP networks than are strictly required for circuit signaling. Even when this is done, there is obviously a limit to the routing capability of the MTP, for instance, to transfer a message between NSP1 and NSP2 in Figure 3.37. This restriction is overcome by the signaling connection control part (SCCP), which we will now discuss.

3.5.4.3. Signaling Connection Control Part

The SCCP was introduced to enhance the capabilities of the MTP in two general ways:

1. to transfer messages between any two Signaling System Number 7 exchanges in the world
2. to support a connection–oriented network service between any two Signaling System Number 7 exchanges in the world

These features are not strictly required for circuit–related signaling but become very important for such purposes as mobile networking, intelligent networking, and support of advanced network management in the evolving ISDN.

Four classes of service are provided by SCCP, two for connectionless services, and two for connection–oriented services (see Q.711):

 0: Basic connectionless class. This transfers unsequenced datagrams, called *unitdata messages,* between SCCP users.
 1: Sequenced connectionless class. This again transfers unitdata messages, but provides sequenced delivery by making use of the signaling link selection (SLS) feature of the MTP.
 2: Basic connection–oriented class. This provides unified procedures for establishment, clearing, and identification of virtual connections, and the transfer of data on them.
 3: Flow control connection–oriented class. This enhances the data transfer capabilities of class 2 with flow control and other features, to offer a full OSI layer 3 network service to its users, allowing support of OSI stacks in the network.

Figure 3.38 illustrates SCCP functionality to provide generalized routing and connection oriented-services. Note that if class 3 is used, the service provided is the equivalent of X.25.

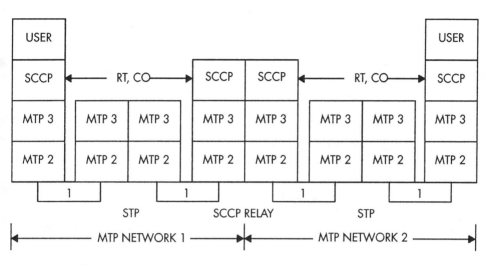

Figure 3.38 SCCP functionality.

Generalized Routing

The SCCP provides a routing based on an address called *global title*, which in its most general form consists of an E.164 number identifying any ISDN exchange (and, indeed, subscriber) in the world. The connectionless service makes use of this as follows. The unitdata message contains a called address in the form of a global title. At the originating exchange SCCP translates this into a DPC, which in the general case is not the final message destination. The MTP transfers the message to the node identified by this DPC; this node typically belongs to more than one MTP network such as ISP1-NSP3 in Figure 3.37. Here SCCP translates to a new DPC and relays the message through the MTP. This relaying continues until the message reaches the final destination node, where SCCP delivers the message to the destination user, identified by the subsystem number (SSN) that identifies the destination SCCP user in a manner reminiscent of the service indicator of the MTP. If routing is impossible at any point, the unitdata is converted into a special unitdata service message, which is returned to the source user.

For Class 0, this is essentially the whole story, with signaling link selection (SLS) randomized at each hop. For Class 1, sequence is maintained by always choosing the same SLS for the same user stream at the source node, and preserving this SLS across the whole path.

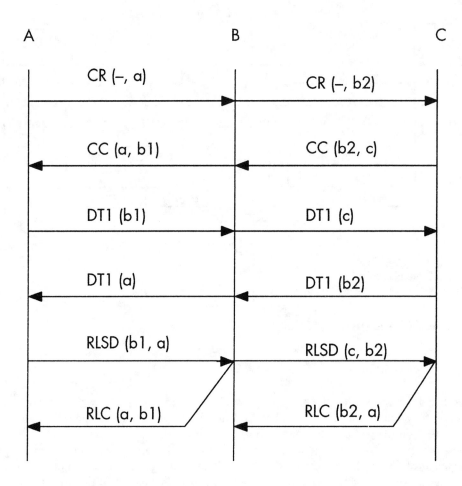

(x, y) = destination reference x, source reference y

Control Data:

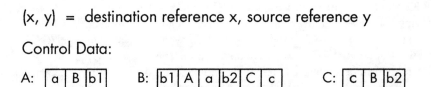

Figure 3.39 Connection establishment.

Connection Service

The basic connection-oriented class 2 of the SCCP enhances generalized routing with procedures for connection identification based on the use of independently assigned local references, a technique shared with the Transport Layer 4 protocol specified in CCITT Recommendation X.224. The technique is also used in commercial correspondence, in which "our reference" and "your reference" are often exchanged at the beginning of a dialogue and subsequently used to identify the transaction in the letter heading.

In SCCP, each node allocates a three–octet reference, which it uses to identify the connection. No structure is specified for this in SCCP, but it will normally be chosen to give most efficient access to control data for the connection within the specific architecture of the node. References are exchanged during connection establishment and are subsequently used to identify the connection, as shown in Figure 3.39, which illustrates a connection between nodes A and C via an SCCP relay point at node B, which chooses two references, one for reception from A and one for reception from C. The SCCP messages used and general procedures are as follows:

CR(–,a)	Connection request message. A chooses reference a and declares this in the CR. Apart from this identification, the CR is similar to a unitdata message, with called address that may be a general global title for routing across the world.
CR(–,b2)	B chooses reference b1 (for A–B) and b2 (for B–C) and declares b2 in the CR, which it routes like unitdata. It stores A and a. C chooses reference c, stores B and b2, and forwards the connection request to its user.
CC(b2,c)	Connection confirmation message. C gets confirmation from its user and declares its reference c in the CC, which it sends to B with B's reference b2.
CC(a,b1)	B stores C and c and declares b1 to A. A stores B and b1 and confirms connection to its user.

The connection control data is now complete and is shown in Figure 3.37.

Routing now proceeds by simple access to control data, as follows:

DT1(b1)	Dataform 1 message. A sends data to B with B's reference. The data may be segmented into several DT1s.
DT1(c)	B uses b1 to access control data, which it uses to relay the DT1 to C with C's reference. C passes the data to its user.
DT1(b2)	Dataform 1 message C sends data to B with B's reference. The data may be segmented into several DT1s.
DT1(a)	B uses b2 to access control data, which it uses to relay the DT1 to A with A's reference. A passes the data to its user.

Release takes place hop–to–hop using the released (RLSD) and release complete (RLC) messages, with both source and destination references included for security. The technique obviously extends to any number of relay points like B, to span any number of MTP networks, and obviously reduces when no relay is needed (A and C on the same MTP network).

The flow control connection-oriented class 3 uses the same technique for connection identification. However, it uses numbered data called dataform 2 (DT2), with acknowledgment (AK) for window rotation, together with messages for expedited data (interrupt in X.25) and reset procedures. Thus SCCP offers a global and general data communication network capability by enhancing the efficient, but localized and specific, capabilities of the MTP.

3.5.4.4. ISDN User Part

The ISUP directly uses the MTP for the purposes of circuit–related signaling. Between any two exchanges, a particular circuit is identified by a circuit identification code (CIC), which follows the MTP routing label in ISUP circuit–related messages. For a given circuit, the ISUP also uses the same signaling link select (SLS) in the MTP routing label, in order to guarantee sequencing. From the point of view of data communication, the CIC acts as a hard-wired "logical channel" enabling direct access to control functions for the circuit. ISUP messages received through the MTP are then distributed based on the OPC (and SSF) in the routing label, together with the CIC. For our purposes, the most important message types used by the ISUP map (in a slightly simplified view, sufficient here) into DSS1 messages are as follows:

IAM (initial address message) maps into SETUP
ACM (address complete message) maps into ALERT
ANM (answer message) maps into CONNECT
REL (release message) maps into DISCONNECT-RELEASE
RLC (release complete message) equivalent of RELEASE COMPLETE
USR (user information message) maps into USER INFORMATION-CONGESTION CONTROL.

Figure 3.40 illustrates the use of these messages for a circuit–switched call between two exchanges.

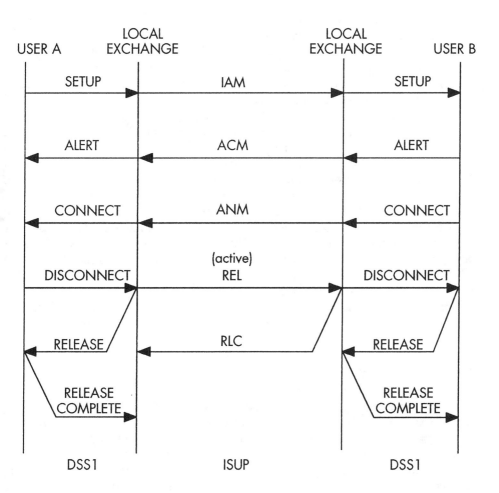

Figure 3.40 Circuit-switched call setup.

3.5.5. User–to–User Signaling in Signaling System Number 7

To complete this chapter, we discuss transfer of user–to–user signaling through Signaling System Number 7.

3.5.5.1. Circuit–Related Signaling

UUS1 is achieved simply by carrying user–to–user information in ISUP signaling messages discussed in the preceding section.

UUS2 and UUS3 can be supported by use of the ISUP USR message to convey the contents of DSS1 USER INFORMATION or CONGESTION CONTROL messages. This method is called *link–by–link*. Its major disadvantage is that the messages must traverse the same path as circuit–related signaling, which is often not the most direct way of sending messages between the end exchanges. This inefficient routing will both increase delays and unnecessarily increase load on MTP signaling links. In addition, it complicates dimensioning of ISUP entities, which must now be capable of relaying user messages, in addition to their normal task of handling circuit–related signaling.

UUS2 and UUS3 can be supported in a more efficient manner by using an SCCP class 2 connection directly between the end–exchanges, giving efficient routing and bypassing intermediate ISUP entities. However, rather than sending a separate SCCP connection request, the source reference and other information is embedded in the ISUP IAM (if the service is invoked at call establishment) or in a special facility request (FAR) message (if the service is invoked after the call is established). This is done to preserve the association between the ISUP controlled circuits and the SCCP connection. This method is called *end–to–end*, or the SCCP method. Once the SCCP connection is established, ISUP USR messages are conveyed in SCCP DT1 messages.

Both methods are defined in Signaling System Number 7. The end–to–end method is preferred, but fallback to the link–by–link method is performed in case SCCP support is not possible end to end.

3.5.5.2. Non–Circuit-Related Signaling

There is as yet no accepted standard for support of USBS in Signaling System Number 7. One approach, used in some countries, is the equivalent of the inefficient link–by–link method for circuit–related signaling. A spare bit in the CIC is set to indicate a "virtual" circuit, and pure ISUP procedures are used to control the nonexistent circuits. Although the approach simplifies signaling procedures, it creates a whole new nonexistent network for which routing tables must be designed, circuits dimensioned and managed, and so on. It requires ISUP signaling entities at each intermediate point and completely ignores the general capabilities of the SCCP discussed earlier.

A much more efficient approach is to establish an SCCP class 2 connection to support the temporary signaling connection between the end exchanges. This makes full use of the general data communication capabilities of the SCCP and does not require definition and dimensioning of nonexistent circuits and a nonexistent network. It shares SCCP functionality with other SCCP users such as management, intelligent network, and mobile network entities, and will automatically benefit from evolution of the signaling network to higher-speed links, possibly carried via ATM mechanisms. The only question with the SCCP method concerns complex signaling. The SCCP itself defines only minimal signaling, sufficient to establish connections between users in exchanges, but without the rich variety of supplementary service signaling defined in ISUP and DSS1. There are basically two simple solutions to this problem:

1. Embed DSS1-like messages (without call reference or channel identification) in the user data of SCCP messages.
2. Embed ISUP-like messages (without circuit identification code) in the user data of SCCP messages.

The first solution has the advantage that it minimizes format changes on the boundary of the Signaling System Number 7 network. However, it does require extension of DSS1 to achieve intranetwork signaling, such as closed user group interlock code, calling and called categories and so on. This extension becomes quite complicated if a rich variety of supplementary services for USBS are offered by the network.

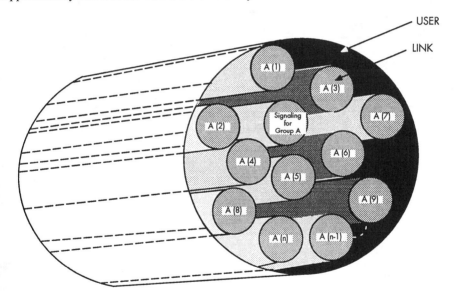

Figure 3.41 Out-of-band signaling.

The second solution requires more format changes on the boundary. However, this interworking between DSS1 and ISUP is well defined, because it is performed for circuit–related signaling. The major advantage of the approach is that ISUP formats and procedures are explicitly designed for intranetwork signaling and can be taken over directly for support of the USBS. Closed user group interlock code, calling and called categories, and so on, are already defined. The rich variety of circuit–related supplementary services associated to circuits can be borrowed where appropriate, without requiring further definition.

International agreement on support of USBS by Signaling System Number 7 would round off the discussion of packetized signaling in this chapter and would provide a unified, integrated, packet-switching capability for the important class of low-speed transactions typified by signaling.

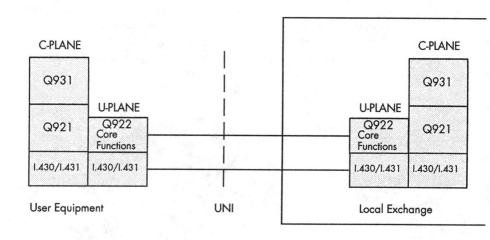

Figure 3.42 User network interface (UNI).

3.6. NEW PACKET MODE

3.6.1. Introduction

The term *new packet mode* (NPM), sometimes also called *additional packet mode bearer service* (APMBS) has been defined for a study within CCITT that covers all possible evolution starting with the current X.31 standard. CCITT Recommendations I.122, I.233, and Q.922 specify a protocol that is based on frame relaying or frame switching. In this protocol the double-call control that is necessary when using X.25 in ISDN can be avoided. A more efficient packet service is defined in which the call control is performed out of band using the techniques of circuit switching for call setup and operation and maintenance. The packets are switched through the network using only the layer 2 frame level.

Two variants are possible :

- frame switching, in which the full functionality of layer 2 is performed by the network
- frame relaying, in which only basic layer 2 services are performed by the network (such as link address conversion, routing, frame sequence check); the other functions (such as flow control, error correction) are left to the user

Frame switching is still a more conventional X.25-like service, because frame relaying puts the emphasis more on performance. It was initially conceived as a multiplexing technique to fully utilize the 1.984 Mb/s of the primary rate access (PRA), but multiplexing speeds up to 45 Mb/s can also be achieved with the same concept. However, by leaving flow control out of the network and by increasing the maximum allowed data rate, new techniques are required to control congestion in the network. Because frame relaying is a more performing protocol than frame switching, primarily frame relaying will be further discussed here.

3.6.2. Frame Relaying Bearer Service

3.6.2.1. *General principles*

The frame relaying bearer service (FRBS) is a new connection-oriented data communication protocol. FRBS is described within CCITT Recommendation I.122. It aims to support a wide range of data applications and data rates from low to high (initially up to 2 Mb/s).

Frame relaying operates entirely within the data link layer and uses the multiplexing possibilities of LAPD. It statistically multiplexes different user data streams at the datalink layer within the same access channel. During the bidirectional transfer of user data, the right sequence of the data is preserved. Each user data stream is called a data link connection (DLC). To identify the different DLCs within the same access channel, each DLC is assigned a label. This label is also called a *datalink connection identifier* (DLCI) and is assigned when a frame relaying call is being established. Permanent logical link connections (PLLC), which are assigned at subscription time, can also be set up.

Frame relaying bearer services have clearly distinct characteristics in the control plane and in the user plane. The control plane functions are responsible for the setup of the connection through the user–to–network interface (UNI). These functions are performed both by the TE and the network. Control plane actions are performed out of band and separate from the user plane actions (see Figure 3.41). Control plane actions use protocols that are the same for all ISDN telecommunication services. The idea is to use procedures (slightly modified Q.921 and Q.931) that are lined up with the procedures for setting up a circuit switched connection.

The user–plane functions are responsible for the transmission of user information through the UNI. The user plane procedures at layer 1 are based on CCITT Recommendations I.430 and I.431. Layer 2 procedures are based only on the core functions of CCITT Recommendation Q.922. These layer 2 procedures allow for the statistical multiplexing of user information. The datalink sublayers of layer 2 are not used. The user network protocols to be used between the TE and the local exchange applicable to FRBS are summarized in Figure 3.42.

To see how frame relay works, assume that application A wants to send a file to application B (see Figure 3.43). Application A starts the communication process by sending a request for session establishment to the transport layer via the presentation and session layers. The transport layer forwards call control information through the ISDN via the D channel using Q.931 procedures. The signaling message is routed through the network and is used to define the virtual path and call parameters that will be used during the data transfer stage. Once the call is established, data is transferred through the network between applications A and B on a hop–by–hop basis, by using the DLCI in the frame header and routing information at each node as determined during call setup.

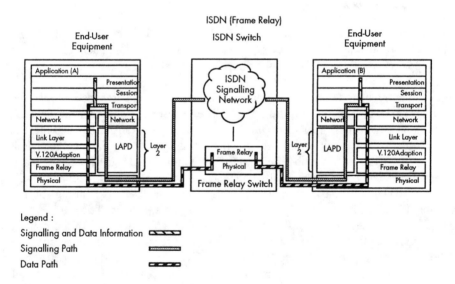

Figure 3.43 Protocol architecture.

The Frame Relaying Network

A frame relaying network will provide efficient communications between various stations on different LANs (such as Ethernet, token ring, FDDI), as shown in Figure 3.44. The access to a frame relaying network is via an interface circuit (IC), such as a bridge or router, plus an appropriate circuit mode facility. Examples of such access facilities may be ISDN channels (D, B, or H channels), leased lines, or MAN bridges.

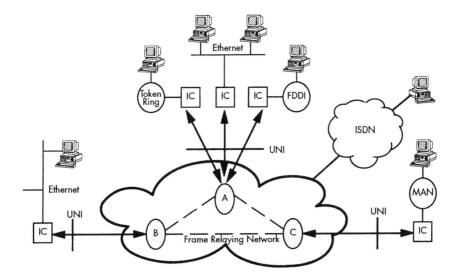

Figure 3.44 Frame relaying (network).

3.6.2.2. Frame Structure in the User Plane

As shown in Figure 3.45, frames are built up by encapsulating layer 2 messages with a two–byte header, a CRC, and flag delimiters. The frame relay header consists of a datalink connection identifier (DLCI) that allows the network to route each frame on a hop–by–hop basis along a virtual path defined either at call setup or at subscription time (in case of PLLC). The start and end flags and the CRC are identical to those used by the protocol HDLC. Procedures for segmentation and reassembly can be used optionally. And when using a terminal adaptor, end–to–end signaling of link errors and control status information can be provided. Figure 3.46 shows the resultant multiplexing scheme in comparison to a 2Mb PCM stream.

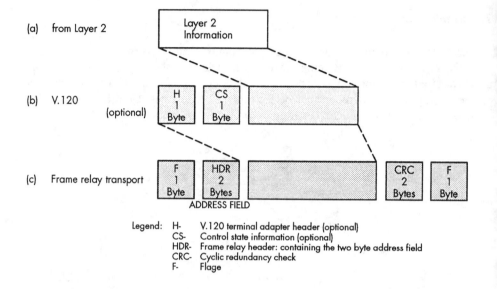

Figure 3.45 Transformation of layer 2 format to frame relay.

Address field. The address field consists of two octets, as illustrated in Figure 3.45. The format of the address field is shown in Figure 3.47. The address field includes: the following bits:

1. address field extension bits (AE)
2. C/R bit
3. explicit network congestion notification bits FECN and BECN and discard eligibility bit DE
4. DLCI subfield (11 bits)

AE and C/R bits have the same meaning as explained in A.2.2.1 for the HDLC protocol, the significance of BECN and FECN is explained in Section 3.6. The DLCI identifies the logical connection, multiplexed within a bearer channel, with which a frame is associated. All frames carried within a particular circuit and having the same DLCI value are associated with the same logical connection. The high- order and low-order parts of the DLCI form a single 11-bit field. The DLCI can thus have a value between 0 and 2047. Table 3.3 shows the range of DLCI values to be used across a FRBS UNI as defined in I.233.

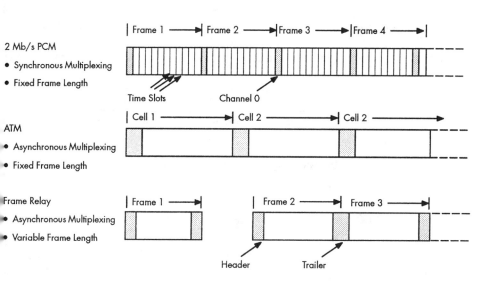

Figure 3.46 Multiplexing schemes.

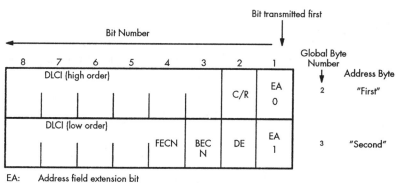

EA: Address field extension bit
C/R: Bit intended to support a command/response idication; the coding is user application specific and is transparent to the network
DLCI: Data Link Connection Identifier Bit d0 is the least significant bit and d10 is the most significant bit.
FECN: Congestion (avoidance) Forward notification bt
BECN: Congestion (avoidance) Backward notification bit
DE: Discard Eligibility bit

Figure 3.47 Address field format.

To enable the support of numerous simultaneously active logical connections on a single FRBS UNI, different DLCI values (having only local significance) are used. For some applications, such as the connection of a host computer to a set of high-speed terminals, the amount of different DLCI values could be close to 2000.

DLCI	Function
0	reserved for in–channel signaling, if required
1-63	reserved for in–channel layer management, if required
64-2015	Assignable for frame-relaying connections
2016-2047	reserved for in–channel layer management, if required

Table 3.3 Two-byte DLCI values.

The layer 2 sublayers. In frame mode bearer services, the layer 2 has been divided into two logical sublayers called the *core functions* for the lower part and the *data link control* for the upper part. A frame-relaying network only implements the core functions (or lower part) of layer 2. Because a frame-relaying network does not implement the upper part of layer 2, whatever datalink control sublayers the users choose to implement end to end are not treated by the frame-relaying network.

The core functions for data transfer are defined in Q.922 as follows:

- frame delimiting, alignment, and transparency
- frame multiplexing or demultiplexing using the DLCI subfield
- inspection of the frame to ensure that it consists of an integer number of bytes prior to zero bit insertion or following zero bit extraction
- detection of transmission errors
- congestion control functions, for reporting congestion to the user
- inspection of the frame to ensure that it is neither too long nor too short

With the above core functions as a basis, a frame-relaying network provides the following attributes in the U–plane:

- full duplex transfer of frames
- preservation of the order of the frames from one UNI to another UNI
- nonduplication of frames
- detection of transmission, format, and operation errors (for example, frames of unknown format)
- transport of the user data contents of a frame transparently; only the frame address and FCS fields may be modified

- no acknowledgment (to the user) of frames within the network
- signaling of the start of congestion

3.6.2.3. Control-Plane Requirements

The C–plane procedures for virtual call control are via out-of-band signaling, wherein one logical link contains the signaling for the other logical links connected to the same user (see Figure 3.41). Layer 2 and layer 3 protocols to be used on the signaling link are derived from Q.921 and Q.931, respectively. The signaling link is established on the D channel. The other links can be on the D, B, or H channels (see Figure 3.48). In the case of permanent virtual circuits, no real-time call establishment is necessary, and all parameters are agreed upon at subscription time.

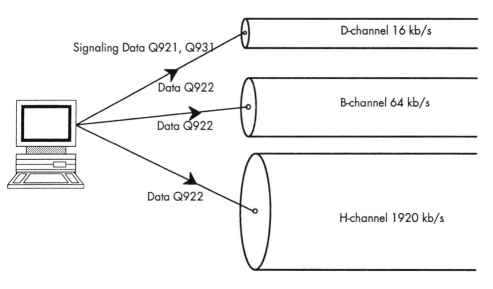

Figure 3.48 Possible channel types for frame relaying.

Access Protocol Extensions for Call Setup

The access protocol (that is, the call setup) for frame services are derived from the actual Q.931 protocol that was mainly defined for circuit-switching services. Preliminary studies have shown that this is feasible. Minor additions are necessary to the existing Q.931, however, because of the typical characteristics of FR-like higher bandwith, out-of-band signaling, and so on.

Following are the most important items to be added:

1. DLCI negotiation
2. access channel negotiation
3. bearer service negotiation
4. call clearing
5. information elements
6. negotiation of parameters

Negotiation of the datalink identifier. Either the terminal dictates the identifier (DLCI) or it is a negotiation between the user and the network. The first option allows the setting of the DLCI to be decoupled from the channel negotiation and is therefore the best choice. It is to be noted that such a choice implies that the DLCI is chosen by the user across its interface in a unique way independent of the channel. The DLCI field could be structured as in Q.921; that is, it could contain an SAPI and TEI point. If this structure is chosen, then TEI extensions may be needed. The possible extension should result in a two-byte TEI field via the field extension mechanism or the addition of a subfield LLI (SAPI–TEI–LLI). Use of an LLI allows, for non D channels, to do some checking in the network about the terminal owner of the channel. This can be interesting for passive bus applications.

Access channel negotiation. It is simpler evolutive to dissociate channel negotiation from datalink negotiation. A channel can be free (when neither frame-mode nor other-mode calls are made to this channel) or busy (with two cases: busy by a nonframe call or busy because a frame-mode call is already made to this channel). At the originating side, the terminal asks for a channel. The network proposes a free channel or a frame channel at which there is still throughput available. At the terminating side, the network proposes a free channel and the terminal chooses; it has the option to propose a busy channel. The network checks the validity, as expressed before, of the channel that was indicated by the terminal in its response to the terminating call.

Bearer service negotiation. The negotiation may be necessary because for some call configuration and some network topology, frame relaying may not be possible at the required level of service quality. In addition, this negotiation brings flexibility because the network has the option to choose the service according to the QOS requirements from the user. The network can also choose the bearer according to its own internal capabilities, which in return gives the call a better chance to be successful while providing the required quality of service. The negotiation can either provide the exclusive bearer capability (BC) (if possible, or clear the call; or provide the preferred BC (if possible) or as alternative BC; or clear the call.

Call clearing (two versus three messages). Data calls are usually cleared via a two-message procedure, which is already defined in some parts of Q.931, for example, user signaling connections. However, the possible use of advice of charge for a frame call may require the use of a three-message procedure because some elements of charging may be evaluated at the access side and have to be received at the clearing initiator extremity, which requires more time. Thus the three-message procedure is necessary as well.

Information elements. Some information elements have to be modified, such as "bearer capability," because of the different possibilities, and "cause," which should reflect errors of data transfer that are not recoverable. Furthermore, new information elements are necessary, such as the frame datalink identifier (FDLCI). The purpose of the FDLCI is to indicate to the network the identity of the LAPD connection that will be used in band to transfer data. The FDLCI information element is set by the calling user to the network in the SETUP message. On the called interface, it is set by the called user to the network in the first returned message, in response to the incoming SETUP.

Besides new information elements, new frame datalink layer parameters are also to be defined; namely all parameters directly affecting the LAPD operation.

Negotiation of parameters. The parameters that are negotiated between users and network correspond to data transfer phase parameters within the network. Negotiation is done toward a lower QOS value. The negotiation is a three-party negotiation: the originator proposes (default is used when no explicit request), and the network may adjust those values and propose the adjusted values to the destination access side. This latter may refuse the call or adjust further the parameters and signal them to the network in its response to the call offering. (No explicit parameter request implies acceptance by the destination of the value.) The network signals back the acceptance of the call with values of parameters indicated only when different from the originator's proposal or default value.

Coordination Between User and Control Plane

Synchronization between out-of-band and in-band for call setup. The function is needed for frame relaying because, when the terminal upper layer has accepted the call via connect response primitive, it can send data to the calling side, but the in–band relation with the calling may not yet be established and will only be established after the end of the OOB signaling. This would result in loss of data. It may be necessary to verify the connection prior to the beginning of the data transfer. This should be accomplished by the two terminals end to end in the user plane.

Synchronization between out-of-band and in-band for release of the call. The release of a frame mode call can be done via OOB signaling with or without prior release in band of the logical link. The release of the logical link or more precisely, the stopping of transfer when done in band prior to OOB signaling, allows a clean separation of transfer allowed and transfer not allowed. By in band, we mean that the terminal, when requested by its upper layers to clear the call, would release at link level the LAPD before issuing OOB signaling to clear the call. The physical channel at the access has to be released when no frame calls are operating over it. The release can be implicit (no indication is given– the channel is cleared at the same time as the last call over it) or explicit (the network gives an indication to the terminal in order to release that channel in a synchronized way).

3.6.3. U-Plane Congestion Control (Reference CCITT I.370)

Because there is no flow control performed by the network, congestion in the data transfer of a frame-relaying network may occur when the traffic exceeds the network dimensioned capacity level or when one of the equipment resources fails. It can also occur at the access due to congestion within the user. The result is performance degradation in terms of throughput, frame loss, and frame delay.

3.6.3.1. General Requirements of Congestion Control Mechanisms

Congestion control mechanisms belong to the data transfer plane (user plane) and are provided partially by the user and partially by the network. The mechanisms have to be developed such that variance in quality of service is equally spread over the users. The network will provide frame relaying resources such as processing capacity, intermediate storage, and transmission links. Proper dimensioning of these resources is the best protection against congestion. However, data transmission networks cannot economically be configured for worst-case conditions. Therefore, appropriate defense mechanisms have to be built. These should react fast, have the adequate granularity, be adaptive and tailored to the circumstances, and allow the users out of forced idleness as soon as the congestion disappears. The user has to implement a dynamic window algorithm, meaning that the time that the data rate is lowered depends on the number of occurrences of congestion per time unit. See Figure 3.49: the terminal lowers the data rate during a longer period of time (t1 = t4 < t2 < t3) when the congestion persists.

3.6.3.2. Levels of Congestion

Figures 3.50 and 3.51 show two levels of congestion in terms of impact on class of service. Point A is the point beyond which the transit delay in the frame relay network increases at a rate faster than the rate at which offered load was increased. This is because the network enters into a mild congestion state. This point is the final point on the curve that the network can guarantee the subscribed class of service. At point B the network begins dropping frames to control the existing level of congestion and prevent additional damage to the network provided services.

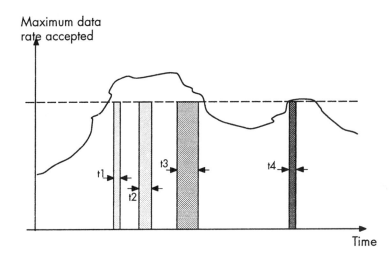

Figure 3.49 Adapted stop-duration.

The end user may perceive the movement from point A to point B without increasing offered load (for example, resource failure or reconfiguration within the network). The different lines in the severe congestion region of Figures 3.50 and 3.51 reflect the fact that different networks may react and degrade differently under severe congestion, depending on the overload capacity of the network resources. The first level of congestion begins at point A. It occurs when the traffic load offered to the network reaches a level beyond which network performance objectives may not be sustained. It is the transition from no congestion in the network to the presence of mild congestion. Point B is the point at which the network begins to discard frames to control the existing level of congestion and prevent additional damage to the network. Points A and B are dynamic values determined by the instantaneous condition of the network resources. These dynamic values are determined relative to the U–plane quality of service objectives to the end user. Network providers may define different values, reflecting different performance objectives (for the support of different grades of services), even within the same network.

Congestion can be controlled by trying to avoid it or by recovering after it has happened. *Congestion avoidance procedures* are activated at point A to prevent congestion from progressing to point B. Congestion avoidance procedures operate around point A and within the regions of mild congestion and severe congestion are activated at point B to prevent congestion that would severely degrade the quality of service delivered by the network. *Congestion recovery procedures* operate around point B and within the region of severe congestion.

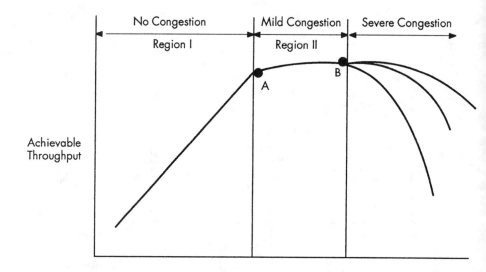

Figure 3.50 Offered load.

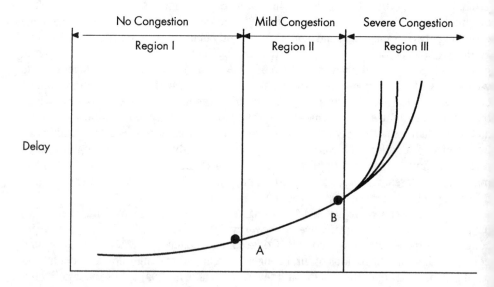

Figure 3.51 Offered load on shared resources.

3.6.3.3. Results of Congestion-Control Mechanisms

Congestion-control mechanisms have to maintain the requested performance levels in terms of throughput, frame delay, and frame loss for each virtual call or permanent virtual circuit. The end–user congestion-control mechanisms have to maintain with minimum deviation the user-perceived quality of service. The network congestion control mechanisms must achieve the following:

- minimize frame discard
- minimize the number of times that quality of service degradation is seen by the end user
- minimize the possibility that one user can monopolize network resources at the expense of other users
- maintain the subscribed quality of service for the user with high probability
- be simple to implement and place little overhead on either user or network
- create minimal additional network traffic
- distribute network resources fairly among users
- limit spread of congestion to other networks and elements within the network
- spread the degradation equally over all users
- operate effectively regardless of the traffic flow in either direction between end users
- have minimal interaction or impact on other systems in the frame-relaying network

Congestion recovery mechanisms (in addition to the above) aim to ensure recovery of the network from a severely congested state.

3.6.3.4. Congestion-Control Mechanisms

Congestion-Control Mechanisms Around Point A

The network generates explicit congestion notification (ECN) to the user beyond region I (point A). In the address field two bits are assigned for this purpose: one bit is assigned to the forward direction (FECN) and the other to the backward notification (BECN). The forward congestion notification is conveyed by setting the FECN bit to 1, the backward congestion notification is conveyed by setting the BECN bit to 1. When there is congestion an ECN is generated in both directions. The network(s) must convey the BECN toward the source user and the FECN towards the destination user, as shown in Figure 3.52. All networks transport this notification without changing it even though they may not generate any.

User Actions When ECN is Received

Whenever the user receives an ECN from the network, it has to lower the rate of information transfer across the UNI immediately. Its behavior as perceived by the network at the UNI is as if it had implemented a dynamic window algorithm per DLCI.

Congestion-Control Mechanisms around Point B

The FR network uses frame discard as an implicit congestion notification mechanism to recover from a state of severe congestion (beyond point B in Figures 3.50 and 3.51).

User Actions When Frame Discard is Detected

Whenever the user detects frames discarded by the network, it must reduce the rate of information transfer across the UNI to allow the network to recover from the state of severe congestion. Its behavior as perceived by the network at the UNI is as if it had implemented a dynamic window algorithm per DLCI. The combined congestion recovery actions of the user and network will bring the congested network back from region III to region I of Figures 3.50 and 3.51.

Network Actions During Abnormal Conditions

The network discards frames received with frame check sequence (FCS) errors without notifying to the sending CPE. The network shall takes no actions as a result of discarding a frame with FCS error.

Network Congestion

The network generates implicit or explicit congestion notification or both to the users during network congestion. It generates explicit congestion avoidance notification whenever it is in a state of mild congestion, and it starts to discard frames whenever it is in a state of severe congestion.

3.6.4. Applicability of Frame Mode Bearer Services for ISDN

3.6.4.1. Teleservices

ISDN frame mode bearer services were initially studied in a bearer service environment; that is, these services evolved out of the need to define an ISDN packet service that would provide a service similar to X.25 in data networks without any of the inherent drawbacks of X.25. In later studies, however, it became increasingly evident that the true potential of ISDN frame mode services lies in their ability to transport a large number of teleservices that do not require the transparent 64Kb/s transport. The most obvious examples of teleservices that would benefit from this mode of transport are videotex, fax, and various types of teleaction teleservices. This list is far from exhaustive.

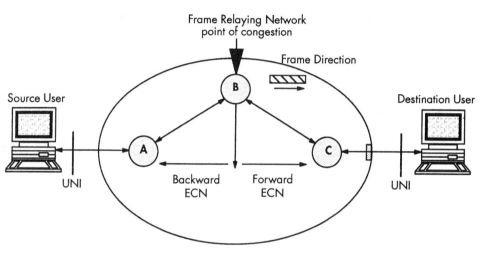

Figure 3.52 Congestion notification.

The decision whether to offer ISDN frame mode services as bearer services or as lower-layer attributes of teleservices depends obviously on the individual network operator. Using frame mode transport as the lower-layer attribute of teleservices has many advantages: the network operator that provides the teleservice is fully aware of the transport requirements of the higher-layer services, the flow characteristics, and the exact need of service data unit acknowledgments. Knowing the application being transported and knowing the characteristics of the end terminals (because the provision of a teleservice, unlike a bearer service, encompasses the characteristics of the end terminal), the network is able to use simple transport mechanisms, with minimum flow control.

The simple rule for deciding when a teleservice may be a candidate for transport by frame techniques may be stated as follows: if the upper layers of the teleservice send and receive blocks of data within upper and lower limits acceptable to the frame mode techniques, and if the quality of service provided by frame transport (error control, delay, congestion protection) is deemed adequate, then one of the two frame mode techniques

(FR or FS) may be used as the lower-layer attribute of the teleservice by the ISDN. In practice, most applications ensure a degree of quality of service at their own layer (flow control, service data unit conformed reception), making it possible to use the most basic of frame techniques (FR) for their needs.

3.6.4.2. Videotex

Because this service is becoming popular in many countries, videotex could be a candidate for use of frame mode transport in an ISDN environment. FR techniques are adequate for the transport of videotex service data units. FR can be introduced as the lower layer for videotex in such a way that the effect on the customer perception of the service is minimal

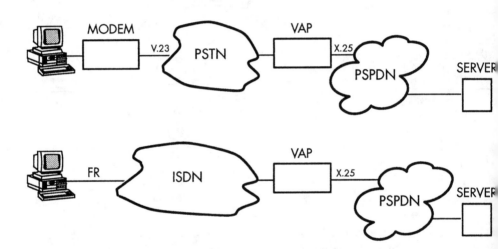

Figure 3.53 Two possible videotex networks.

A normal videotex terminal today uses the voice band of a telephone connection to transmit and receive bytes, based on the V.23 modem recommendation. The network's videotex access point (VAP) then assembles or disassembles the received bytes into X.25 packets, which are sent to dedicated servers connected to the PSPDN. The voice band -V.23 mechanism can be replaced by an FR link (established for each call), which transports data blocks on n–bytes each to and from the VAP (see Figure 3.53). The server connected to the VAP are not modified in any way. The main consequences of this solution is the implementation of FR ports on the VAP and the creation of a new generation of videotex terminals using FR instead of V.23. The main investment both in the VAP and in customer terminals is related to the upper layer functions presentation, dealing with dedicated function keys, and so on. From a service point of view, the big advantages of this solution are, apart from a considerable improvement in the error control, the ability of both the user and the network to better use the transmission capacity of available circuits due to the low rate multiplex possibilities.

3.6.5. Interworking Between NPM and Other Protocols

The I.500 series of CCITT describes several possible interworking arrangements between frame mode bearer services (FMBS) and existing or new protocols:

- interworking between FMBS and X.25/X.31 (see Figure 3.54)
- interworking of LANS and FMBS
- interworking between two alternative FMBS: frame relaying and frame switching
- interworking between FMBS and circuit-switched services
- interworking between FMBS and ATM

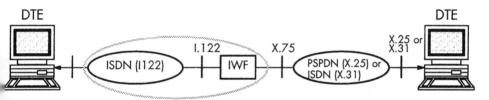

Interworking function (IWF) logically belongs to ISDN (I.122)

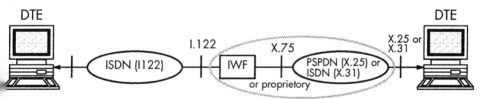

Interworking function (IWF) logically belongs to PSPDN (X.25) or ISDN (X.31)

Figure 3.54 Interworking FMBS and PSPDN (X.25) or ISDN (X.31).

The extent of the list indicates that there is a potential proliferation problem when all these possible interworkings have to be provided in many national variants. From a supplier point of view, a threat of a multiplication factor to be applied to the already vast number of functions, attributes, and services drives at more cautious position. By explaining only two of the above interworkings, it will be illustrated that the complexity is sufficiently high to try to avoid providing all the possible variants if it is not absolutely necessary.

Interworking Between FMBS and ISDN Circuit Switching (see Figure 3.55)

As in the minimum integration scenario of X.25 in ISDN, a circuit-switched call has to be setup to the remote frame handler using the Q.931 protocol. A virtual call is thereafter established using the frame mode call procedures in band as if the TE were a subscriber of the frame mode network. At first glance a very similar structure as for minimum integration of X.25 in ISDN has been adopted, but from a development point of view if such an interworking has to be implemented, then the possibly higher bandwidth (up to 1920 kb/s), the unavailability of X.25 as a packet protocol, and the double call setup will make it a totally new network with its own live, own operation and maintenance.

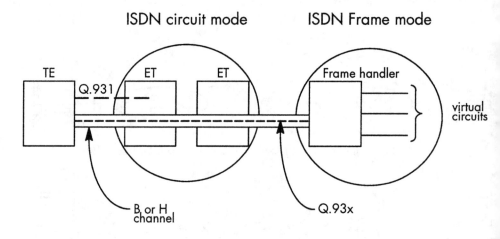

Figure 3.55 Interworking between FMBS and ISDN circuit switching

Interworking Between FMBS and ATM (see Figure 3.56)

The interworking function plays the role of originating and terminating subscriber because this is set up in two parts:

1. from the FMBS TE to the interworking function
2. from the interworking function to the ATM TE

This technique is called *call control mapping*. During the call setup the user plane parameters (such as throughput and maximum frame size) have to be negotiated between the two networks through the interworking function.

3.6.6. Advantages of Frame Relaying

The typical characteristics of frame-relaying techniques have some clear advantages, which come from the establishment of the call in one step, the clear separation between signaling and data transfer and the multiplexing of user data at the lowest level.

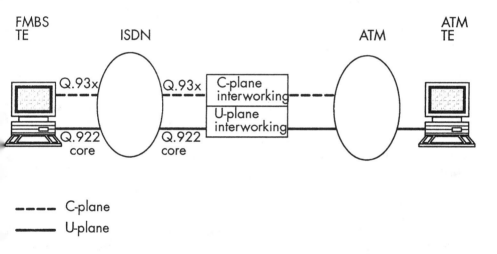

Figure 3.56 Interworking between FMBS and ATM.

3.6.6.1. Advantages in building an exchange

Equipment efficiency

For the same investment in switching equipment a frame relay switch can handle five to ten times more data than a packet switch of the early generation. Because of the redundancy in the protocol a packet switch has to treat an average of eight frames per packet of user data. A frame relay mode handles only four frames for the same user data because much less overhead information has to be treated, but also less processing time is required per frame, resulting in the above-mentioned factor of five to ten.

The current generation of packet-switching systems, combined with a good quality of transmission network, achieves a better performance, but even then an improvement factor of two can be achieved with frame relaying. However, in order to reach this improvement the right balance has to be achieved between throughput and processing time. If the bitrate becomes very fast, then the processing capacity becomes a limiting factor. If the processing capacity is not adequate, then the transit delay increases and queues build up, thus effectively decreasing the transmission speed. Thus the best possible compromise has to be formed between bitrate, processing capacity, and buffer length. All three elements have an important influence on the product cost, so they should be selected in the optimum proportion to each other.

Furthermore, PSPDNs, by implementing frame relaying techniques as part of their network, can achieve similar performance improvements. It is to be expected that they will do so in order to be able to continue to compete.

Lining Up with Existing Functions

Because the signaling system, both on the user access and the network level, can be lined up with circuit switching, a saving can be achieved in the development cost. The same is valid for the supplementary services. The long list of already existing supplementary services for ISDN subscribers using circuit switching can be reused for frame relay subscribers.

Simplifications

Because different quality of service requirements may be valid for signaling than for data transfer, and because both are kept separate in the network, the design can be adequately proportioned. In addition, the network can be provided with the required services at the points in the network at which where they are essential. This means that in certain nodes in the network a lower protocol intervention level could be tolerated, for example a layer 1 through connection without extra functions on top of it, or in other words, a pure circuit connection. Finally, the frame relay protocol can also be used between modules of the exchange because it is a simple protocol.

3.6.6.2. *Advantages for the user*

Because of the simplicity of the protocol and the lining up with existing protocols, the user sees the following advantages:

- the possibility to benefit from a higher performance
- shorter delay times
- the same signaling systems across various bearer services
- the same supplementary services simplifying the terminal and the usage of them
- the same lower-layer mechanisms between bearer services
- option for users to select functions on top of the basic protocols
- easy support of connection-oriented as well as connectionless types of services
- possibilities for point-to-multipoint connections
- an expected lower cost in comparison to X.25 because of the higher network efficiency

3.6.6.3. Advantages for the administration

Because frame relaying fully integrates a data communication technique within the ISDN network, a simpler operation and maintenance structure is possible than when two networks have to be supervised. Higher revenues will be generated by this network because of the higher transmission speeds.

3.7. MAN (METROPOLITAN AREA NETWORK)

3.7.1. Introduction

Distributed queue dual bus (DQDB) systems, defined by IEEE, and switched multimegabit data services (SMDS), defined by Bellcore, implement a high-speed (140 Mbps) packet network. This dual bus network is used primarily for communication between LANs, which connect local terminals, workstations, and computers, and between PABXs, effectively forming a wide area network (WAN). Because the traffic generated by LANs is bursty and of the datagram type, the WAN switches packets on a connectionless basis. This means that for transfer of data between two end users, no connection has to be set up or released; each data packet carries the full destination address and control information necessary for routing the packet. In this way the network occupied only when there is data to be transmitted. The dual bus system, although better suited for packetized data traffic, also allows synchronous services (n x 64 kbps channels) to be routed semipermanently via the WAN network toward other networks (such as PSTN and ISDN) or to other users (PABX) on the dual bus, such that an efficient high-0speed access is presented to the user. This network is intended primarily for covering business needs in metropolitan areas (hence the name metropolitan area network, or MAN), with later provision for interconnection of large cities. The MAN is the first step toward broadband networks, because it is targeted to satisfy the demand of business customers and will offer an interface for later connection to ATM.

3.7.2. Other High-Speed Networks

The LAN market has also seen an evolution on the subject of high-speed and integrated services. The most commonly known and successful example is fiber distributed data interface (FDDI) defined by ANSI as a 100 Mbps fiber–based LAN system. The main features through which the DQDB system distinguishes itself from FDDI are its ability to use standard transmission systems (G703 and SDH/SONET), the random spacing of network elements along the shared medium, the segmented transfer and ATM compatibility of the protocol, and the optimization of the system for use in public networks by its point-to-point bus capability (for access links), the privacy and security features, and its network management system. In what follows we will describe only the DQBD system.

3.7.3. MAN Basic Switching Principle

3.7.3.1. Medium Access Control and Standardization

Before explaining the network structure and service provisioning, it is necessary to clarify the core switching principle: a high-speed dual bus structure, to which multiple stations have access, in order to exchange packets with each other (see Figure 3.57). The bus forms a shared communication medium between the stations; this principle is also used in LAN techniques: Ethernet (IEEE 802.3), Token Bus (IEEE 802.4), and Token Ring (IEEE 802.5). However, while the medium access control (MAC) protocol used by the stations on these LANs suffers from rather high overhead, and therefore allows only partial bus (or ring) load, the DQDB protocol used allows nearly 100 percent loading of the shared medium with user data. In practice, this load will not be allowed, because the access queue delay per station would be intolerably high, and because a certain degree of fairness in bandwidth sharing has to be kept; still a significant improvement over conventional protocols is obtained.

The original MAC protocol has been accepted by the IEEE as the basis for MAN protocol standardization; this standards process has added extra functionality to it, together with ATM compatibility, and is now commonly known as the IEEE 802.6 DQDB standard. In Europe, the ETSI group for MAN issues has also accepted the DQDB as the basis for MAN standardization.

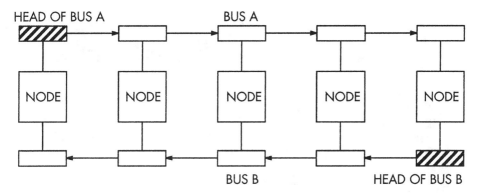

Figure 3.57 MAN dual bus structure.

3.7.3.2. The DQDB MAC Principle

In order to describe the DQDB principle first with a simple analogy: the dual bus system can also be represented by a railway system in which two stations along the railway are interconnected via a telecommunication path e.g. a telex which is the analogy for one bus and by the railway, which is the other bus (see Figure 3.58). A user who wants to send two packets sends a message to the previous user(s) to reserve space on the bus, just as the chief of the railway station sends a message to the previous station to reserve two places on the next train that is going to arrive. These two passengers have priority over other passengers arriving later to their station. In this way, whoever comes first is served first, independent of the station at which the train is taken. This priority system allows the train to be occupied up to a very high percentage without increasing the risk for congestion. In a more technical way the explanation goes like this:

- The operation of and access to a "dual bus" can be explained with the help of Figure 3.59.
- The communication medium consists of two contradirectional buses A and B.
- Access nodes are connected to the bus via an OR gate and via reclocking logic for write access.
- On the buses, a frame structure is present, consisting of "n" slots of 53 bytes, "n" being dependent on the bus bitrate.
- A slot is divided into 5 header bytes and 48 payload bytes (see Figure 3.60).

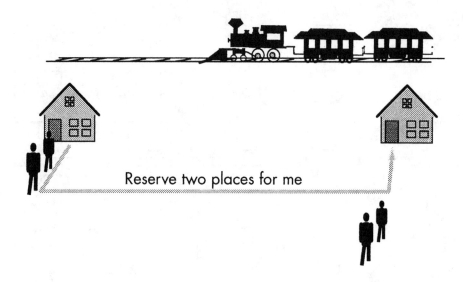

Figure 3.58 DQDB principle in real life.

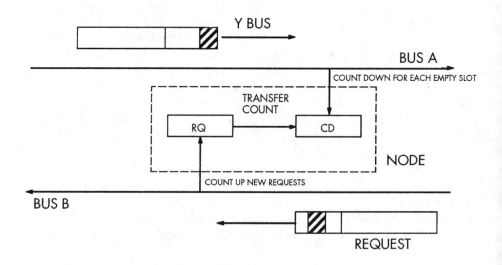

Figure 3.59 The basic DQDB algorithm.

In contrast to ATM, however, in which every service is packetized, the DQDB cells can be used either for asynchronous (packet) services or for synchronous (STM) or isochronous services. For this reason the DQDB frame system is synchronized to 125 μs or a multiple of 125 μs. Slots used for packet access are called QA (queued arbitrated), those allocated for isochronous services are called PA (permanently assigned). Slots for isochronous service are semipermanently assigned to nodes by network management, and the remaining slots can be used by all nodes as a shared bandwidth resource for packet transfer. This frame structure is generated by the first node on the bus (head of bus A and head of bus B). Because the bus works with segments of packets, there is a segmenting and reassembly (SAR) procedure required for variable length packet transfer (see the next paragraph). By using segments, finer granularity in bandwidth occupation is obtained, lowering the average access delay; buffer allocation is also simplified.

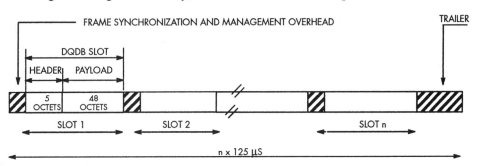

Figure 3.60 MAN frame structure.

Figure 3.59 illustrates the operation of the basic DQDB algorithm: the protocol uses two bits in the header of a DQDB slot, BUSY and REQUEST, to control access to the slots on the forward bus. An identical, but independent arrangement applies for access to the opposite bus. The BUSY bit indicates that a slot is already occupied by a station; the REQUEST bit is issued on the B bus when a station wants to transmit a segment on the A bus, and vice versa. The REQ bit serves as an indicator to the upstream nodes that an additional segment is in the queue. The REQ bit is issued on the distributed to all nodes, and each node keeps track of the number of segments queued downstream from itself by counting the REQ bits in its REQ counter (RQ). The counter is decreased each time an empty slot passes on the forward bus (the empty segments are used by one of the downstream queued segments,) whereby the RQ counter keeps a record of the number of segments queued downstream. Returning to the analogy with the railway system: the head of the railway station knows that all stations at which the train will arrive later have requested a certain amount of seats in the train, says ten. Furthermore, he himself has sent forward to all stations at which the train arrives earlier that he wants free seats also says five. So, the head of the railway station will count until he has seen ten empty places pass by, assuming that all ten were reserved earlier than his five. Then he will allocate his five passengers, knowing that all stations in front of him will leave them free, because they were reserved earlier than all remaining requests.

When a node has a segment for access to the bus, it transfers the current value of the RQ counter to a count down (CD) counter and resets the RQ. This action loads the CD with the number of downstream segments queued. The CD is decreased for every empty slot passing on the forward bus, and the node transmits its segment in the first empty slot after CD reaches 0. While waiting for access, any new REQ bits received from the other bus are added to the RQ. This way, a single FIFO queue is established among all nodes for access to the bus. In addition, the protocol provides priority by using separate REQ bits and counters per priority level; the counters of lower level will count requests of lower and higher levels, whereas the CD counter will be increased for REQ bits received at higher priority levels.

3.7.3.3. DQDB Protocol Stack

The process of conveying a user datagram of variable length onto a DQDB is shown in Figure 3.61. The user data is handed over to the DQDB MAC layer with the parameters source address, destination address, quality of service, and data. Before segmentation, the convergence sublayer adds the necessary protocol information to the data unit: begin and end tags (to detect cell loss) and address information and trailer. This data unit is called IMPDU (initial MAC protocol data unit). Afterward, the IMPDU is segmented into parts fitting the payload of a DQDB cell. Four bytes of the payload are used for segment type (ST), message identifier (MID), length, and CRC fields. The ST field serves to indicate if the first segment of a datagram is received, a subsequent segment or the last segment; the MID field is used to indicate to what datagram the cell belongs (cells of different datagrams can be multiplexed on the DQDB); the length is used for indicating the amount of data in the cell.

3.7.4. Network Structure

In contrast to the conventional LAN configurations, the DQDB links can be configured as point to point, open bus or looped configuration (see Figure 3.62). The looped configuration appears similar to a ring, but data does not pass through the head of the bus; the reason for looping is not only topology, but also to add redundancy: a "self-healing" mechanism allows each node on the loop to function as head of the bus, so that a break in the loop would be healed by positioning the head or end of the bus adjacent to the break and by interconnecting the original head or end of the bus.

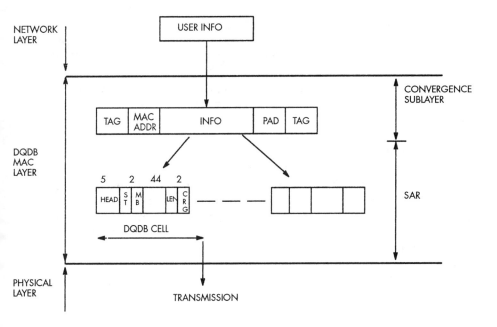

Figure 3.61 User datagram structure.

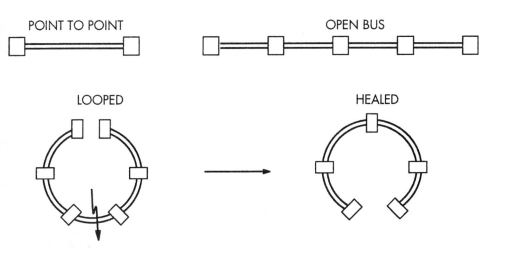

Figure 3.62 Open bus or looped bus.

Because the bus also carries isochronous slots, its frame structure must be synchronized to the PSTN clock if connectivity with this network is to be assured; therefore a synchronization signal is to be inserted at a node (any) on each DQDB in the MAN.

Typically, the MAN is formed by interconnection of high-speed "backbone" subnetworks (see Figure 3.63), which in turn interconnect smaller areas covered by lower-speed subnetworks to which subscribers are connected (customer access networks, or CAN). Currently, the subnetworks work either at 34, 45 or 140 Mbps, whereas the CAN operates at 34 or 45 Mbps; a 2 Mbps CAN version is also possible. The maximum length of a subnetwork, with random station spacing, is about 150 km. A maximum of 200-250 nodes can be allocated on the bus; these maxima are a function of loop delay affecting the DQDB protocol and of allowable jitter on the bus. The maximum length of a DQDB bus with only the end–stations equipped (for example, to interconnect MANs or customers to MANs) is equal to the maximum length allowed by the used transmission system. Subnetworks are interconnected using router network elements, routing data on a per cell basis.

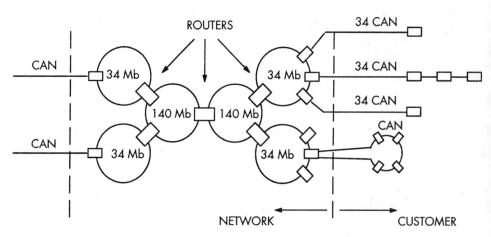

Figure 3.63 A MAN network.

This type of network can be deployed in private as well in public environments; the difference lies in the connection of customer equipment to the subnetwork backbone. For private applications (such as a high-speed LAN or interconnection of low-speed LANs within the customers premises), the user equipment can be connected directly to the subnetwork (see Figure 3.64); there is no need for authentication procedures, no securing the data from being observed by other user equipment. In this case, the MAN equipment that interconnects customer applications with the backbone is called customer network interface unit (CNIU). This is the most economical network structure, because it requires minimal equipment. On the other hand, in cases in which the MAN is used in a public environment, privacy, security, and authentication has to be assured; therefore, the equipment of each individual customer is connected via a separate network element to the backbone (see Figure 3.65). In this configuration, the public network element terminating the customer access network (CAN) is the edge gateway (EGW), and the network element connecting the customer applications to the access line is the customer gateway (CGW); the CGW is here part of the CPE (customer premises equipment). The CAN is also formed by a DQDB link: this enables multiple CGW, remote form each other, to be connected to the public network (see Figure 3.62). In addition, the use of E164 addressing and the NMC facilities assure the applicability in a public network.

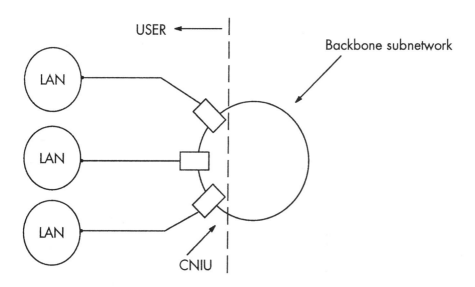

Figure 3.64 User access to the MAN.

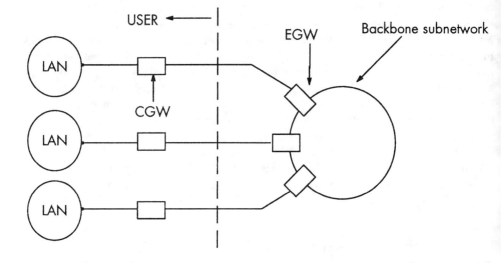

Figure 3.65 User access via a gateway.

3.7.5. Services

3.7.5.1. Connectionless Packet Switching

The primary end service offered by the MAN is LAN interconnection. LAN interconnection preserves the customer's investment in LAN terminals and software; no changes are required in the end user equipment. This is accomplished by providing a MAN bridge, for interconnecting LANs of the same type, or a MAN gateway, for dissimilar LAN interconnection (see Figures 3.66 and 3.67). The bridge translates the destination MAC address into a destination E164 number, "encapsulates" the whole LAN packet in a network packet, and routes it via the MAN to the destination (see Figure 3.66). A gateway (sometimes called a *router*) can work with LANs of different MAC addressing procedures by using an internet protocol (such as DOD IP, Decnet or XNS) on top of the heterogeneous networks used; the sending terminal gives a destination IP address, the gateway translates it into a destination E164 MAN address, and finally the destination gateway translates the IP address into the destination MAC address (see Figure 3.67).

3.7.6. Q3 and SMDS Protocol

3.7.6.1. Introduction

The network bearer service offered by the MAN is connectionless packet-switching, provided by using the "Q3" network layer protocol at the EQW/CNIU network elements. This protocol runs on top of the DQDB subnetwork protocol (layer 2), explained in section 3.7.3. The complete protocol stack functionality of LAN interconnection on MAN is shown in Figures 3.66 and 3.67. The Q3 protocol is proprietary and provides the routing of datagrams on E164 address basis, together with screening, authentication, charging, policing, group addressing and QOS selection features; a maximum of 7K bytes can be carried in one datagram. However, because these procedures are used at the access of a public network, at a certain point in time standardization is required.

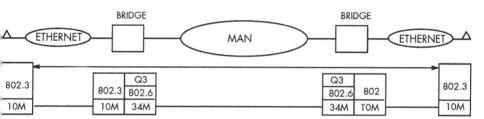

Figure 3.66 Network and protocol structure.

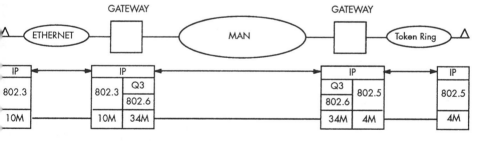

Figure 3.67 Network and protocol structure with gateway.

3.7.6.2. U.S. Standardization

In the United States, the process of MAN network layer standardization is currently being conducted by Bellcore in the form of technical advisories; TA772, 773, 774, 775 describe the switched multimegabit data service, or SMDS, to be used on 802.6 DQDB access links, called here subscriber network interface (SNI) the SMDS provides a superset of the "Q3" protocol. In SMDS terminology, the geographical MAN area is covered by MAN switching systems (MSS; see Figure 3.68), which can use either centralized or distributed technology. The MAN on DQDB can also be used either as distributed or centralized switch. The MSS could be supplied by different vendors, so again, a standard interexchange and inter–MAN interface is required. Work is currently underway to produce advisories for the interswitching system interface (ISSI) in addition to the SNI standard; this interface will most probably be in line with the ATM NNI.

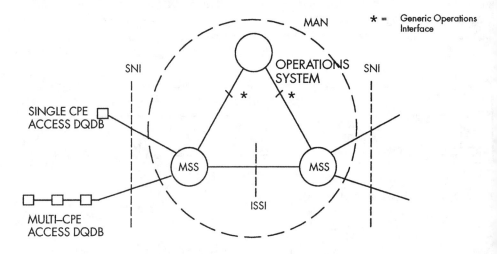

Figure 3.68 The MAN switching system.

3.7.6.3. Europe Situation

Because of the fact that SMDS is very focused on the U.S. market, its regulations, and its numbering plan, we expect that the MAN standardization process for Europe (which is currently ongoing in ETSI) will yield its own or modified specification. In the meantime the Q3 protocol is being upgraded to the full functionality of SMDS, so that SMDS–like service can be offered on the European market.

3.7.6.4. Isochronous Services

With the introduction of metropolitan area networks, business customers who are connected to the network via 34 or 45 Mbps access links are given the opportunity to use

64 kbps channels within this bandwidth using PA slots (e.g., for PABX interconnection and connection to the PSTN). The MAN acts as a crossconnect, routing the channels on a semipermanent basis (see Figure 3.69). In addition, two kbps CAS signaling channels are routed along with the 64 kbps channel; while the routing for D channel signaling for ISPBX is a pending development.

Note that the isochronous services are not packetized as in ATM; 125 ms synchronization and delay characteristics are preserved such that this solution can be seen as a cost-effective "early" access integration, awaiting for ATM terminals, PABX, and networks.

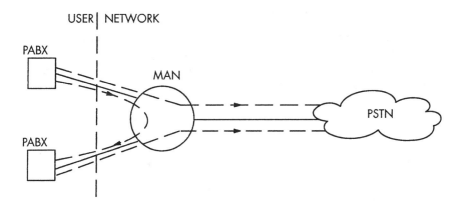

Figure 3.69 The MAN as a crossconnect.

3.7.7. Product Evolution

The evolution of the MAN product from today's features can be divided into new applications, new network features, network management evolution, and migration toward and final integration into the broadband ISDN.

3.7.7.1. Application Evolution

A relatively short-term development is the FDDI bridge and gateway. Initially, performance will be on the order of magnitude of the present LAN bridges and gateways, evolving toward high-performance FDDI interconnectivity.

Packet relay applications will enable customers with X.25 terminals to be connected via the integrated MAN access toward the PSPDN. Using this relay function, ISPBX could also be interconnected, using semipermanent D channel signaling routing (on L2 basis). Limited video for business applications are envisioned, using low bitrate (1-5 Mbps) signals and simplified connection procedures. Low-cost 2 Mbps DQDB access will bring MAN service within reach of most small business users. Finally, SONET (SDH) DQDB access, for high–end business customers, will provide the performance and a step in the direction toward B–ISDN.

3.7.7.2. Network Feature Evolution

The interconnection of private MAN to public MAN as a short-term evolution will enable large business customers to deploy their own networks, numbering plan, and management system, while maintaining a gateway toward the public network. The advent of SMDS at SNI and ISSI, and its European version, will be taken into account in the MAN in order to provide universal interconnectivity with standard terminals and exchanges. Extension of MAN subnetworks is envisioned by using ATM interconnect nodes (see Figure 3.70).

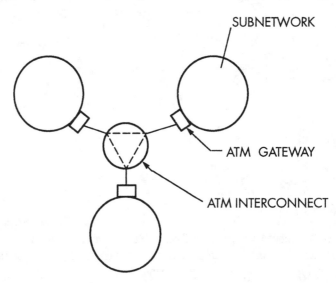

Figure 3.70 MAN interconnection through ATM.

3.7.7.3. Evolution Toward B-ISDN

The evolution toward B–ISDN will be in stages; one of the first phases will be the introduction of ATM interconnect nodes at the transit (intersubnetwork) level.

The next phase will be the upgrading of these nodes to switching capacity for other broadband services, eventually also including MAN–type services (see Figure 3.71). In this case, the MAN can be regarded as a concentrator and switch for connectionless services in the interface between MAN and B-ISDN. At the access, the fiber–to–the–business (which might be installed at MAN service introduction), will be used to provide access to NB and BB services, in a minimum integration fashion (see Figure 3.72).

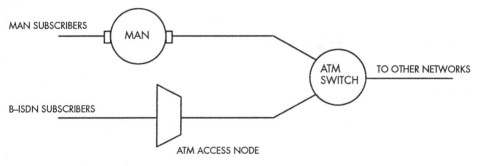

Figure 3.71 The MAN as a concentrator in an ATM network.

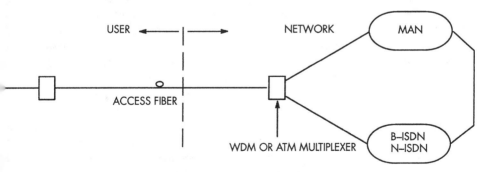

Figure 3.72 Fiber-to-the-business through MAN.

The reader may whish to consult the following references :

- "Generic System Requirements in Support of Switched Multimegabit Data Service," Bellcore TA–TSY–000772, Issue 3, October 1989.
- "Distributed Queue Dual Bus (DQDB) Metropolitan Area Network (MAN)," proposed standard IEEE 802.6/D11.
- "The Alcatel MAN Overview Manual," October 1989.
- "Alcatel MAN: High Speed Metropolitan Area Network". Electrical Communication magazine volume 64 Number 2/3 1990.

3.8. ASYNCHRONOUS TRANSFER MODE

3.8.1. How to Handle Higher Bit Rates

The 1980s saw tremendous advances in transmission technology, mainly in the deployment of optic fibers, characterized by:

- huge inexpensive bandwidth; the order of Gbps on cables became cheaper than traditional coaxial copper
- essentially perfect quality; increasingly small bit error rates were achieved, without the burst errors characteristic of metallic media

In parallel advances in semiconductor technology made possible the design of switches capable of handling these huge bandwidths in a flexible manner. Techniques for dividing up this bandwidth between information streams, and switching these streams, were widely discussed in this period as part of the studies related to the broadband ISDN. (Broadband refers to bandwidths above the narrowband standard of 64 kbps, although the narrowband ISDN is capable of circuit switching limited multiples of this standard.)

The general term *"transfer mode"* is defined by I.113 to refer to aspects covering transmission, multiplexing, and switching in telecommunications networks. Discussion ranged between the two broad transfer mode categories of circuit and packet.

After a great deal of study, the decision was made in CCITT to adopt a fast packet-switching technique known as asynchronous transfer mode (ATM) to flexibly utilize bandwidth up to 150 or even 600 Mbps. Beyond these speeds, a circuit-switched approach called synchronous transfer mode is used. This is also capable of carrying the traditional 64 kbps streams of the narrowband ISDN, multiplexed to higher-order bit rates.

Synchronous Transfer Mode

This is the traditional circuit–switched approach. It basically involves dividing up the available bitstream into fixed–length frames consisting of a recognizable header, called a *frame alignment* or *sychronization pattern*, followed by a sequence of n groups of b bits each. Each group corresponds to a separate continuous bit rate connection: if the frame repeats every T seconds, then the bit rate per connection is b/T, and the total bit rate is $(s + nb)/T$, where s is the length in bits of the synchronization pattern plus other overheads related to management of the bitstream.

For example, the most well–known framing is performed at the primary rates of 1.544 or 2.048 Mbps. Both schemes have $T = 125$ microseconds and $b = 8$ bits, giving a per-connection bit rate of 64 kbps, the narrowband ISDN standard. In both schemes the bits associated with a certain group (called a *channel*) are contiguous. The first was introduced earliest, and uses $s = 1$ and $n = 24$, whereas the second uses $s = 8$ and $n = 31$. The second is illustrated in Figure 3.73.

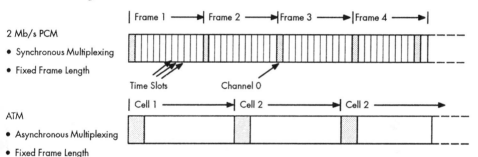

Figure 3.73 2Mb Frame structure versus cell structure.

Note that the word *frame* in this context is distinct from the HDLC meaning. Both divide up or "frame" the bitstream, but they do so in different manners, and they process them completely differently. For some reason, synchronous framing is put at layer 1 of the OSI model, where as HDLC framing is put at layer 2.

After detection of the alignment pattern, called *frame acquisition*, switching is very simple. For a given connection, the received bits on an incoming channel are assembled and transferred to an outgoing channel, and the process is repeated every T seconds. This is the basic function of a narrowband ISDN switch. Note that the most important requirement for high quality is that clocks be identical in frequency over the whole network; distribution of high-quality clocks from a central cesium clock is a major feature of network design.

Because of its simplicity, such synchronous switching can be performed at very high speeds. At the highest speeds available on optic fibers synchronous transfer mode is therefore used, and the bitstream structuring is specified in the synchronous digital hierarchy (SDH) of CCITT based on the SONET specification from the United States. SDH is important because it takes advantage of the universal clock to allow streams to be combined and split (multiplexed and demultiplexed) in a very flexible and simple manner. Earlier multiplexing schemes beyond first order are designed to handle streams whose real bit rate may be rather different from the nominal speed. They include complex mechanisms to allow for this tolerance, which is in any case not acceptable in the ISDN.

Asynchronous Transfer Mode

Although simple, synchronous transfer is very rigid and inflexible; a choice must be made on how to divide up the bandwidth, and once this is done, the decision is frozen in both transmission and switching equipment. As an extremely important example, take the all-pervasive 64 kbps standard in the ISDN. This is based on the transfer of speech at 8000 8–bit samples per second, known as *pulse code modulation*. Even for speech at a constant bit rate, this now seems too rigid:

- There exists a standard scheme for 32 kbps speech, called ADPCM;
- An acceptable quality is achievable for speech at 16 kbps, or even lower, as shown by the digital mobile phone network GSM.

In addition, the use of a 64 kbps channel in both directions cannot take advantage of the fact that, in most conversations, one party is speaking while the other is listening.

On private networks, particularly in the United States, a more flexible and economical use of leased bandwidth has been achieved by packetizing speech. Typically, ADPCM and silence recognition are used to reduce the required bandwidth by a factor of up to four; one scheme uses the 1.544 Mbps framing discussed above but replaces the 24 channels by 24 octets of packetized speech plus identification. Another advantage of these schemes is that computer data as well as speech can be carried in these packets, making for an extremely flexible use of available bandwidth.

For the bandwidths available in the broadband ISDN, one must also consider information flows such as high-quality audio (compact disc requires 1.4 Mbps without compression) and various video qualities (up to 150 Mbps without compression). The arguments for packetization then become overwhelming:

- Even if a service uses a constant bit rate, continuous advances in technology may reduce the required bit rate over the course of time.
- These advances can be of benefit only if the network can switch streams of any bandwidth, rather than adopting a standard minimum rate.
- Further advances in technology often result in a variable bit rate or asymmetrical bit rate requirement for most services. This "burstiness" is obvious for computer

communications and was the reason for the adoption of packet-switching. The same is true for sophisticated speech encoding; and video often alternates between semistatic images and rapidly changing images.
- These advances can be of benefit only if the network can switch streams with variable bandwidth requirements.
- New services that we cannot yet identify may need to use the capabilities of the broadband ISDN—neither their bandwidth nor the burstiness of their traffic can be predicted.

Such arguments have been familiar for a long time. However, it is only in the last few years that technology has advanced to the stage at which one can contemplate packet, rather than circuit, switching at such blistering speeds. Such switching must obviously be performed in hardware; it must therefore be as simple as possible. In addition, it must be capable of carrying any type of information, not just computer data.

Asynchronous transfer mode is the simplest possible form of packet-switching as we shall see. Even then, it is much more complex than synchronous transfer mode, essentially because of its statistical, rather than rigid, capabilities. Rather than taking a simple group of bits from a fixed position in a repeated incoming stream and transferring them to a reserved fixed position in a repeated outgoing stream, a switch must absorb a larger incoming group in a block, interpret the label, and transfer the block to a queue for an outgoing stream; it must monitor the flow on each stream, both for accounting purposes and to limit the user to a statistically specified usage; it must also monitor for congestion for interaction with resource allocation control software. In addition, for constant bit rate services, terminal equipment must be capable of handling jitter (the statistical fluctuation in cross network delay) in order to correctly reproduce the continuous stream packetized at the source. Thus the extreme flexibility of fast packet-switching comes at a price in switch and terminal complexity. Such complexity is possible only with technologies that were unimaginable when X.25 was defined.

3.8.2. The ATM Technique

ATM divides up the bitstream into fixed length blocks of 53 octets called *cells*, each consisting of a five-octet header and 48 octets of information. The general cell structure is shown in Figure 3.74.

Cells are transmitted continuously. When no information is to be sent, idle cells are sent, identified by a special preassigned header of nearly all zeros. Special headers consisting mainly of zeros are also reserved for special purposes, such as operations and maintenance.

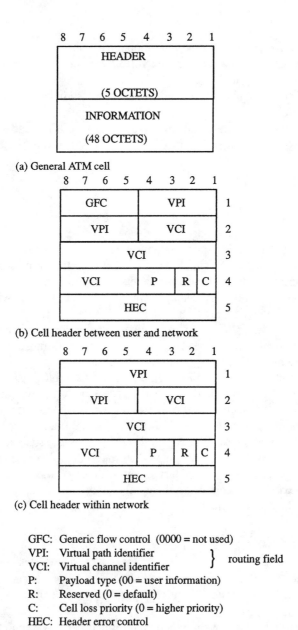

Figure 3.74 ATM cell description.

3.8.2.1. ATM Cell Delineation (Compare HDLC Delimitation)

This is an interesting technique. It essentially finds and locks onto the repeated cell structure by searching for an error-free header pattern (see the next section). Cell information is scrambled to avoid spurious generation of such a pattern. Note that once the cells have been "acquired," reception can operate in a periodic manner every T seconds, where T is a fixed cell transmission time.

3.8.2.2. ATM Error Detection and Correction (Compare HDLC Error Detection)

Every header contains a one-octet header error control (HEC). This is a cyclic code capable of detecting and correcting a single bit error in the header and detecting multiple bit errors. Multiple bit errors, and single bit errors in cells following a corrected cell with a single bit error, cause cell discard. If errors persist, then a cell delineation search is reinitiated. Note that no error control is performed on the information carried in the cell.

3.8.2.3. ATM Routing (Compare HDLC Address and X.25 Logical Channels)

This is achieved in a connection–oriented manner. Each header contains a routing header that identifies the connection on each bitstream. For each connection, cells are switched from an incoming bitstream (called a *multiplex* in ATM) to an outgoing bitsteam, and the routing header is changed accordingly. Note that the use of different identifiers for each direction of flow is under discussion at the time of this writing.

An interesting feature is the decomposition of the routing header into two distinct fields: the virtual path identifier (VPI) and the virtual channel identifier (VCI). This allows for a hierarchical structuring of the network:

- The virtual path permits global routing of all the virtual circuits that belong to it: it is usually seen as a semipermanent identifier allowing routing through a low–level network of multiplexers and cross–connects.
- The virtual circuit within a virtual path identifies the distinct connection, it is usually seen as a dynamic call identifier, treated by high–level switches or servers reached by virtual paths.

This concept allows, for instance, for private network users to be interconnected in a semipermanent but flexible manner by virtual paths, while using virtual channels to multiplex end–to–end traffic. Other virtual paths could connect them to public servers for dynamic call handling or high-speed datagram services. It also allows for similar families of equipment to be used for multiplexing, cross–connecting, and dynamic switching. The only difference lies in the way in which the hardware is controlled.

3.8.2.4. ATM Control

The header has various other fields that allow for extra features to be built on top of the basic ATM routing. Most of these are under further study at the time of this writing, although the cell loss priority is worthy of mention. This feature allows a user to submit more expensive high-priority cells or cheaper lower-priority cells. The lower-priority cells are discarded by the network in case of congestion, giving some elasticity in this case. Possible uses of priority might be to layer the end–to–end stream for, say, video, into a basic, high-priority stream defining a picture of a certain quality, enhanced by a lower-priority stream giving higher quality. This reduces costs to the user, while defining a minimum service to be maintained at all times.

3.8.2.5. ATM Information

The ATM cell contains 48 octets of unprotected user information. The number 48 was chosen as a compromise between the Europeans, who wanted 32 octets, and the Americans and Japanese, who wanted 64 octets. The smaller number was probably desired to limit requirements for echo control equipment for packetized speech. The larger number reduces the overheads consumed by the header; for areas in which satellite telephony is widespread, echo control equipment is often included as a matter of course. For more information about how the ATM technique compares with more traditional packet handling techniques please read section A.6 of Appendix A.

Chapter 4

Description of an Implementation

A description of an ISDN network with the emphasis on packet switching follows. It is assumed that the basic concepts of ISDN are understood.

The distribution of functions is assumed to be like those in a distributed control system. Only the modules needed to perform minimum integration packet-switching functions according to ETSI (STC NA2) are described. However, as said before, the forms of packet switching such as the NPM described in section 3.6 could be performed equally well with the same system structure, changing only firmware and software.

4.1. THE ISDN CHIP SET

The necessary ISDN interfaces can be built by using a limited set of integrated circuits at different places (see Figure 4.1).

ISDN Chip Set Partitioning

The chip set has to repeat the CCITT defined interfaces, but in addition, standardization of chips is more easier if an interchip interface V* is standardized as well. The chip set consists of five integrated circuits covering the basic ISDN functions:

- two–wire UIC (U interface circuit)
- four–wire SIC (S interface circuit)
- ILC (ISDN link controller) for circuit and packet-switching layer 2 functions

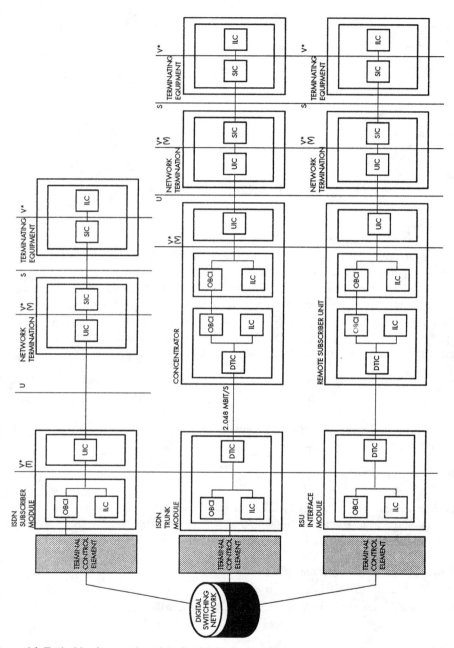

Figure 4.1 Typical implementation of the five VLSI circuits that make up the ISDN chip set.

- OBCI (on–board controller interface) for interfacing the exchange modules to the digital switching network
- DTIC (digital trunk interface circuit) for 2 Mbit/s trunk connections

Figure 4.1 shows a typical implementation of these VLSI circuits in the various ISDN modules for the exchange sides, network termination, and end user equipment.

Because each VLSI circuit in the chip set operates at only one layer of the OSI communication network model, it can be used wherever its functions are required. The UIC and SIC operate at layer 1 of the OSI model, which deals with the physical and electrical characteristics of the link over which data is transported. Both the ILC and OBCI operate at layer 2, which establishes a reliable error–free connection on the physical link defined by layer 1, using HDLC (high-level data link control) protocols for packet switching and CCITT Number 7 common channel signaling. As an example of the flexibility of this approach, the ILC can be used, either for D channel signaling functions or for a 64 kbit/s packet–switched channel. The ILC can also support layer 2 functions in CCITT Number 7 signaling applications and primary rate access with a minimum of additional circuitry.

V^* Interface (Interchip Interface)

The V^* interface supports data transfer, communication at layers 1 and 2, and the transfer of maintenance information within the digital loop. It is a four–wire interface using bidirectional data, clock, and frame signals. The interface can operate in either a bus or chip–to–chip mode. In the bus mode, which is used at the exchange side (V reference point), the data rate is 2 Mbit/s and the clock rate is 4 Mbit/s; in the chip–to–chip mode, used at the terminal side (T reference point), the data rate is 256 kHz and the clock rate is 512 kHz. The four–byte information is transferred in bursts (bus mode) or continuously (chip–to–chip mode). As shown in Figure 4.2, this configuration contains the following:

- two 64 kbit/s B channels for speech or data
- one 16 kbit/s D channel for signaling
- activation and deactivation status bits
- transparent maintenance channel
- monitor enable bit
- eight–bit monitoring channel

U Interface Transceiver Chip (UIC)

The UIC is a layer 1 full duplex interface circuit for two–wire, 144 kbit/s basis access units. Its key functions and characteristics are as follows:

- two- to four-wire hybrid based on echo cancellation for line lengths up to 8 km on 0.6 mm wires
- V* interface compatibility
- activation and deactivation at layer 1 resulting in a reliable physical link setup with a bit error rate of 10-6 after a maximum setup time of 150 ms

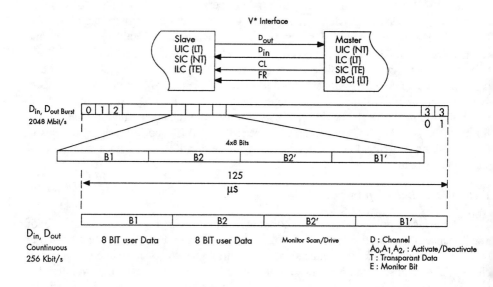

Figure 4.2 Configuration of the V* interface.

ISDN Link Controller

The ILC (ISDN link controller) performs multiplexing and demultiplexing for the B and D channels. Its key building blocks and features are as follows:

- V* interface compatibility
- two HDLC formatters per receive and transmit port, programmable for 64 kbit/s or 16 kbit/s depending on whether packet data is sent on the B or D channel, respectively
- on-board direct memory access controller for packet transfer
- capacity to be programmed for basic access, primary rate access, or CCITT Number 7 common channel signaling applications

S Interface Circuit

The SIC includes the master and slave functions for network termination and terminal equipment applications that are terminated by the V^* interface. It also supports activation and deactivation procedures on incoming and outgoing calls, as well as signal recovery (clock and frame) in the master and slave operation modes.

On-Board Controller Interface

The OBCI performs the function of front–end control element (FCE). The FCE consists of an OBCI (on–board controller interface) and an OBC (on–board controller). The OBCI is an intelligent custom VLSI circuit that provides a control and transmission interface between the network terminals (analog and digital), the processor of the module and OBC. Its main task is to switch any incoming channel to any outgoing channel, including those on its own port, in a transparent way. The FCE is controlled by the OBC firmware. The OBC controls the terminals and switching of the OBCI. The FCE firmware can receive messages from and transmit messages to the software. Basically the FCE performs layer 2 functions, and the modules software performs layer 3 (or higher) functions.

Digital Trunk Interface Circuit

The digital trunk is the interface between various 2048 kbit/s (30/31 channel) digital trunk formats and the transmission format. It switches 64 kbit/s channels between the trunk interface and the OBCI or OBC. In addition it synchronizes the received digital signal with the local clock, converts the HDB3 (or AMI) coded signal to binary (and vice versa), handles multislot connections ($n \times 64$ kbit/s), and detects certain alarm conditions (loss of frame alignment). The HDB3 part consists of the conventional interfacing functions required to convert an HDB3 (or AMI) coded signal to a binary signal that is compatible with the logic circuits, and vice versa. Optionally it provides CRC4 (cyclic redundancy check 4) generation and detection in the trunk transmission format; this can be enabled or disabled by an OBC command.

4.2. ISDN MODULES

The modules needed to perform packet-switching functions are those at the subscriber side (see Figure 4.3):

- the ISDN subscriber module (ISM), or
- the ISDN module for remote access (the ISDN remote subscriber unit (IRSU)), or
- the ISDN private exchange (ISPBX)

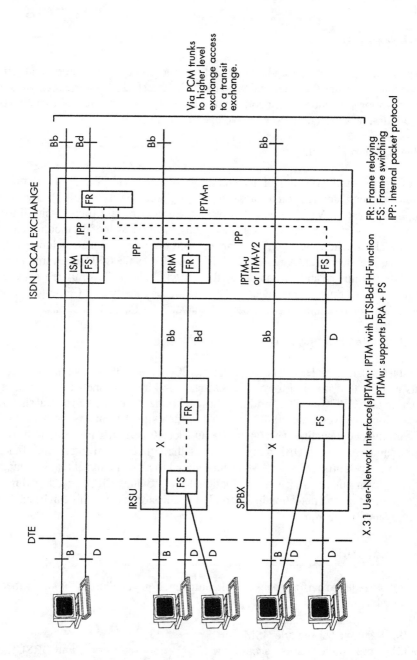

Figure 4.3 Local exchange configurations.

and the modules at the network side :

- IRSU interface module (IRIM)
- ISPBX interface module (IPTMu)
- network interface (IPTMn)

Together those modules perform the basic functions for setup and supervision of packet-switching calls.

The functions to be performed are as follows:

1. B channel services (X.31 Case A/B)
   - semipermanent Bb channels (established by operator command)
   - dynamic Bb channels (established by using signaling procedures) in line with ETSI Recommendations for DSS1

2. D channel services (X.31 Case B)
   - semipermanent datalink and permanent logical link (PLL) access to the packet handler (PH), both via semipermanent Bd channels (established by operator command).
   - on-demand switched datalink and dynamic semipermanent virtual link access to and from the PH established as layer 2 data on on-demand switched Bd channels (established by signaling procedures at first virtual call) and 'dynamic semipermanent Bd channels (established by signaling procedures at subscription time)

With the ETSI STC NA2 packet-switching strategy, all X.25 services are provided by an external packet handler (PH) and are not provided by the ISDN network.

4.2.1. Performance Requirements

An ISM is expected to be capable of handling the packet mode traffic per subscriber as shown in Table 4.1. The table provides traffic figures for a whole ISM (64 BA) based on the CCITT assumptions for a realistic mixture of light, medium, and heavy data users. The table shows calls per second (CAPS) for circuit-switched connections and for packet-switched connections.

Traffic assumptions	A: Light User	B: Medium User	C: Heavy User
CAPH/Subscriber	2	10	50
Packets/s/Subscriber	0.01	0.05	0.25
Packets/Call	18	18	18
Bits/s/Subscriber	10	50	250

Table 4.1 ISM handling of packet mode traffic.

4.2.2. ISM

The ISM must meet the following requirements for implementation of the ETSI packet mode bearer service:

- B channel access to the PH according to X.31 Case A. This requirement however, does not affect the ISM. The subscriber or TA must dial the ISDN number of an ISDN access connected to a port of the packet network. The connection goes transparently through the ISM.
- B channel access to the PH according to X.31 Case B. This requires that the ISM for outgoing calls (to the PH) recognizes in the subscribers SETUP the bearer capability "packet mode" instead of an ISDN number. With that information the ISM has to request the establishment of a Bb channel through the ISDN network to the specific ISDN exchange, which for the ISDN area (to which the calling exchange belongs) interfaces to the PH via primary rate access. For the opposite direction, the PH has to establish a Bb channel to the called subscriber by "dialing" its ISDN number according to CCITT E.164. Once the Bb channel is switched between PH and ISDN subscriber (or is established semipermanently), LAPB on layer 2 and X.25 on layer 3 of this physical channel is used to establish one or several virtual connections between terminal and PH. For these the ISM and the whole connection through the ISDN will be transparent.
- D channel access to PH according to X.31 Case B. This requires that first a Bd channel between the frame handler in the ISDN local exchange and the PH be established. This is, however, not the ISM's responsibility, but is left to the IPTMn and the PH. Bd channels can also be available on a semipermanent basis. The ISM's responsibility is to support establishment of layer 2 links (LAPD) between subscribers and the ET–OBC and to prolong them by IPP across the DSN to the frame handler (FH) in the network-side IPTM (IPTMn). The IPTM–FH prolongs the data links further to the PH.

In the data phase, frames are forwarded on these layer 2 sections between subscriber terminal or adapter and the PH in both directions. The ETSI specifications for packet mode services in the D channel optionally request the ISDN to support either full layer 2 frame switching or just layer 2 frame relaying. Most administrations prefer layer 2 frame switching. Thus the ISM–OBC has to perform full (acknowledged) frame switching, whereas the FH in the IPTM supports just an (unacknowledged) frame-relaying function between ISM and PH.

Figure 4.3 provides an overview of this function split between ISM and frame handler in the IPTMn, and shows the relation to other subscriber-side modules (IRSU and ISPBX).

4.2.2.1. Detailed functions

The ISM functions are performed by the functional units (see Figure 4.4.):

- line termination unit
- exchange termination unit
- terminal control unit

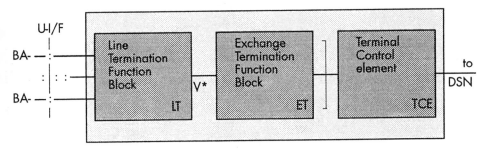

Figure 4.4 ISM module for basic access, functional overview.

4.2.2.2. Terminal Control Element (TCE)

This is the part of the module that contains a processor and its memory. Here resides the software that performs the layer 3 functions.

4.2.2.3. Exchange Termination

The ET function block cuts through the B channels for up to eight basic accesses between the line termination (LT), and the terminal control element, and handles and controls the D channel protocol.

Support of D channel packet mode bearer service as specified by ETSI is alo a specific task of the ET function block. If data packets are transmitted, the ET has to protect them against data violation, that might occur during transmission on the line or in the network. To solve these major tasks the ETC–PBA is set up from appropriate subunits (see Figure 4.5):

- on–board controller (OBC)
- on–board controller interface (OBCI)
- D channel protocol unit
- packet protection unit
- retiming and monitor channel handling unit (RETMO)

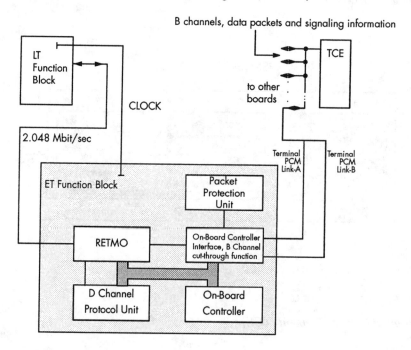

Figure 4.5 ET function block architecture of ISTB-PBA.

On–Board Controller

The on–board controller (OBC) is a front–end microprocessor, made up of the following:

- a 16-bit microprocessor
- memory area maximum.
    - 256 k 16-bit EPROM
    - two 128 k 8-bit SRAMs

    The OBC tasks are as follows:

- initialization, maintenance, and management of the four ILCs for the D channel protocol unit
- support of the D channel signaling protocol
- support of the D channel packet mode by performing the frame-switching function
- use of IPP layer 2 for direct communication with frame multiplexer in IPTMn
- cooperation with the packet protection unit or calculation of CRC–checksum to ensure IPP packet validity
- support for test of the subscriber lines and for error detection and diagnostic routines for line terminating PBAs
- maintenance functions (loop commands, continuous supervision of layer 1, statistical evaluation of frame errors)

D Channel Protocol Unit

The D channel protocol unit has to support the handling of the HDLC protocol on D channels between BA subscriber and ET–OBC. The ET could provide for a B channel protocol unit to support B channels for HDLC transmission as a provision for B channel packet-switching. However, because the ETSI–defined packet mode bearer service does not require L2 handling of B channel packet mode in ISDN, this protocol unity is omitted. The D channel protocol unit is responsible for handling layer 1 and a part of layer 2 of the D channel protocol. It performs the following functions:

1. HDLC controller and formatter function for the D channel protocol of up to eight subscribers according to ETSI specifications

    - full duplex operation
    - automatic zero insertion and deletion
    - address field recognition
    - FCS generation and checking

- interframe time fill
- abort detection and generation

2. serial interface to receive and transmit the D channel maintenance and control information from and to the LT function block.
3. interface function from and to the on–board controller for this information.

It is also a task of the D channel protocol unit to support the activate-deactivate-monitor-maintenance-control information flow between line termination and on–board controller.

The D channel protocol unit takes into account the requirements for error detection in diagnostic routines. In the D channel protocol unit, one HDLC–VLSI — the ISDN link controller (ILC) — supports the D channels of two BA subscribers. The D channel protocol unit in a fully equipped ET unit is made up of four 4 ILCs interfacing between the microprocessor bus and the RETMO. These four ILCs can support eight BAs for circuit switching. The ILCs forming the D channel protocol unit will also be used for support of D channel packet mode, if a subscriber connected to the ILC has subscribed to and is using this service.

Packet Protection Unit

The packet protection unit (PPU) in the exchange termination is a component, that is used to support packet-switching between OBC and the network (or between OBC and TCE, if IPP is used for OBC–TCE communication) by protecting the packets against bit errors during transfer. This is achieved by applying the cyclic redundancy check method, using the polynomial $X16 + X12 + X5 + 1$.

4.2.2.4. Line Termination (LT) (see Figure 4.6)

The LT function handles only layer 1 functions. It is therefore not influenced by the requirements of handling packet-switching functions in the ISM. It can be used for both circuit and packet-switching.

Timing and Clock Functions

Timing for the line termination functions is derived from the systems clock.

U–Interface Function

For provision of the transmission function on the LT (U– to V^*–interface adaption), a VLSI customer circuit (U-interface circuit, or UIC) is used.

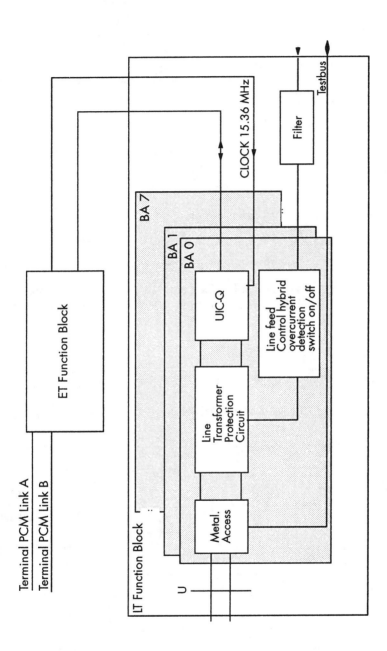

Figure 4.6 LT functional block.

Line Feeding and Overvoltage or Overcurrent Protection

The LT comprises circuits for line feeding and overvoltage protection. The subscriber line feeding is enabled and disabled individually by a command from the ET.

Operation and Maintenance

The LT supports the maintenance functions. It provides line-testing functions including loop back facility (loop 1) at the U–interface, frame error monitoring at the U–interface and loss of synchronization indication.

4.2.3. IPTM (ISDN Packet Trunk Module)

4.2.3.1. User Side (see Figure 4.7)

At the user side we have the packet-handling capabilities of IPTM, according to the X.31 Recommendations (Blue Book) for interfacing a PRA with B channel circuit switching and B and D packet-switching.

4.2.3.2. Network Side (see Figure 4.7)

At the network side the following packet-handling functions are possible:

- IPTM for multiplexing packet data received from the user side onto a Bd channel (ETSI standard)
- IPTM implementing the packet handler interface (ETSI standard PHI)

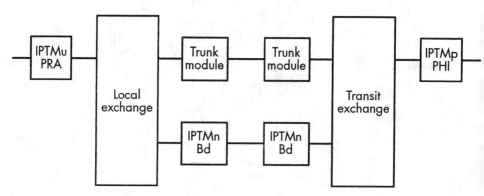

Figure 4.7 IPTM application overview.

4.2.3.3. IPTM at the ISDN User-Network Interface (IPTMu)

The integrated packet trunk module (IPTM) type IPTMu is a module providing primary rate access (PRA) for ISDN subscriber to ISDN networks. Subscribers at PRA level interface with the IPTMu at the CCITT V2M reference point (see Figure 4.8). The module variant enabling this access type has to provide ISDN-access for ISPBX subscribers within the area of an ISDN Exchange. One ISDN PABX can be connected to one or more IPTMs or to one ISM.

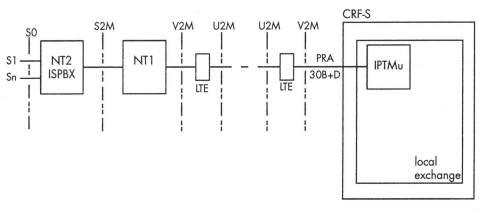

Figure 4.8 IPTM PRA reference configuration.

The ISDN D channel (point-to-point) will be supported (that is, with no multidrop) for signaling only. To meet local security and traffic requirements, more than one access may be provided to an ISPBX.

The IPTMu implements both the control plane and the user plane functionality. It handles the signaling carried by the D64 Channel of the PRA. It also supports circuit-switched traffic over the B channels and packet-switched data from B and D channels.

Packet-Switched Services on the B Channel (see Figures 4.9. and 4.10)

This service is similar to the circuit-switched service in the sense that the packet traffic is not terminated at the IPTMu but is only transiting. The IPTMu has to control the user's B channel and set up a semipermanent B connection or an on-demand B channel toward one or more packet handlers. These B channels carrying packet data will exit the local exchange as Bb channels in any type of trunk module at the network side.

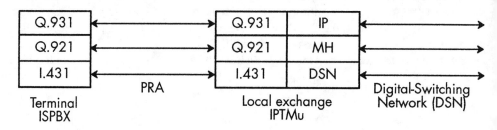

MH: message handler
IP: internal protocol

Figure 4.9 IPTMu C-plane.

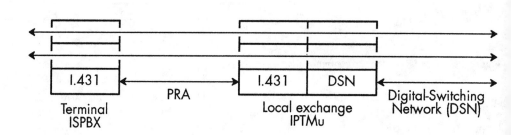

Figure 4.10 IPTMu U-plane.

Packet-switched services on the D channel (see Figures 4.11 and 4.12)

The D channel carries signaling (DSS1) and packet datalinks. This means that the C plane and U plane use the same physical link. The local exchange has to perform frame handling for the D channel. This function is divided into a frame-switching (FS) part at the user side (IPTMu, ISM, or IRSU) and a frame-relaying part at the network side (IPTMn). In the U plane the network provides frame switching for the D channel packet-switching functions. Frames are discarded when any of the following conditions apply:

- They do not consist of an integer number of octets prior to zero bit insertion or after zero bit extraction.
- They are too short or too long.
- They have a faulty FCS.
- They carry an unknown DLCI.
- They cannot be routed further due to internal network considerations.
- Other possible conditions exist, such as policing functions.

Sublayer 2.3, consisting of unacknowledged and acknowledged protocols, and sublayer 2.4, consisting of a link monitor for activation and deactivation, are also handled.

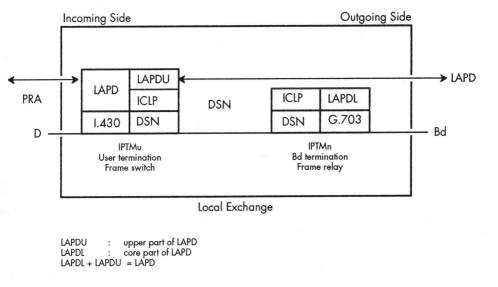

Figure 4.11 Protocol stack for frame handler.

The Q.921E Recommendation implies for the IPTM that additional address information has to be added in order to allow multiplexing of all D channel packet datalinks of several subscribers onto one channel (Bd) toward the packet handler. The module software is not involved with frame switching; only the front-end processor of the IPTMu is active. Frames, with an end-to-end meaning (for example, I-frames) that are received on the D64 channel are forwarded by the FS of the IPTMu via ICLP to the Bd channel found in its control block.

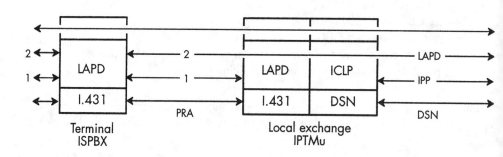

Figure 4.12 IPTMU U-plane for D channel packet switching.

4.2.3.4. IPTM at the ISDN Network-Network Interface (IPTMn)

The IPTMn is located in the local exchange and is complementary to the IPTMu described in the previous section. The IPTMn provides only the U-plane functionality needed for D channel packet-switching; that is, it concentrates packets received from the IPTMu via the DSN onto up to four Bd channels. It performs the frame-relaying part of the FH of the local exchange (see Figure 4.13).

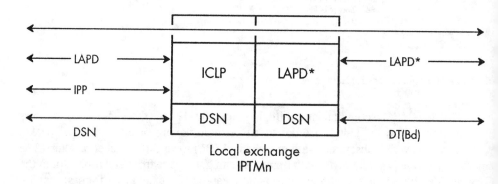

Figure 4.13 IPTMn U-plane for D channel packet switching.

Packet-Switched Services on the B Channel

The IPTMn being fully transparent, performs the function of a standard trunk module.

Packet-Switched Services on the D Channel

The IPTMn terminates up to four Bd channels at the network side of the local ISDN exchange.

4.2.3.5. IPTM at the ISDN Packet Handler Interface (IPTMp)

The packet handler (PH) is connected to the ISDN via the packet handler interface (PHI). For the ISDN the PH is at the subscriber level, and access is performed at the primary rate (PRA). Normal DSS1 signaling applies. As a result of communality, both IPTMu and IPTMp are made out of the same firmware modules.

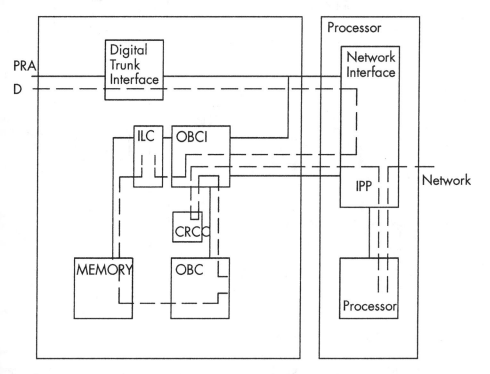

Figure 4.14 D channel signaling (control plane).

Packet-Switched Services on the B Channel

Bd channels from the local exchanges transit through this IPTMp toward the PH. IPTMp is transparent for these packets.

Packet-Switched Services on the D channel

Bd channels originating at the level of the local exchanges transit through the IPTMp toward the PH, and no further multiplexation is done.

As an example, we will examine the IPTMu data flow. Both control plane (C) and user plane (U) are implemented in this IPTM. The signal flow in the control plane is shown in Figure 4.14, and signals in the user plane are shown in Figure 4.15.

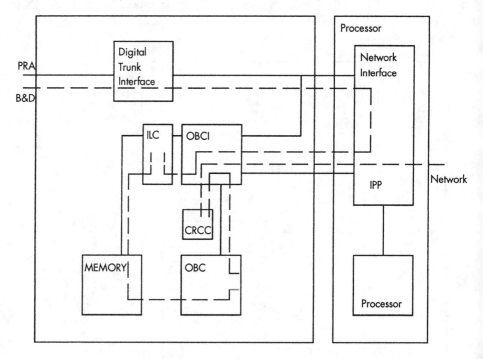

Figure 4.15 B and D channel packet data (user plane).

In the control plane the D channel passes the digital trunk interface and is directed via the network interface and OBCI (on-board controller interface) toward the HDLC formatter in the ILC (ISDN link controller), in which low-level layer 2 functions are performed. From there further handling is done in the OBC (layer 2 and lower layer 3). Higher-layer functions are performed by the processor software, which receives its information using the IPP protocol. The user plane implements a similar signal flow except that it does not pass through the processor and instead goes immediately with the IPP across the DSN.

4.2.4. IRSU (ISDN Remote Subscriber Unit)

Functionally, the IRSU can be viewed as a remote ISM, but in order to connect the lines remotely trunk modules are placed in the middle of it (see Figure 4.16). The ISDN subscribers are connected to exactly the same interface as in the ISM (that is, the LT and ET functions). The other essential component of the IRSU is the DT, which connects it to two PCM links. A simplified block diagram is shown in Figure 4.16. In the parent exchange the PCM links are terminated by ISDN remote interface module (IRIM). The IRIM consists of a DT and a module control unit (MCU).

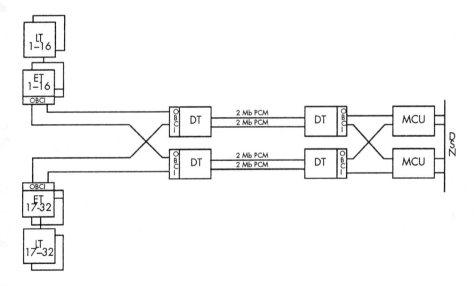

Figure 4.16 IRSU-IRIM overview diagram.

The tasks that are performed in the LT-ET function in the IRSU are exactly the same as for the ISM, including the frame formatting into IPP. Again the IPP works end to end, from the OBC in the ET to the OBC on the IPTMn on the other side of DSN, and the IPP packets are protected by the CRCCs. The necessary addressing information is fetched from the IRIM–MCU at call setup. The IPP packets containing the frames to be transmitted across the DSN will be concentrated on one or more dedicated channels for conveying packet traffic on the PCM link between IRSU and IRIM. More explanation is not required because all other information can be found in the ISM description. However, because the IRSU structure is more complex than the ISM structure, the principles of routing D channels is best explained here.

As shown in Figure 4.3, frame-switching and frame-relaying functions are required at several points in the network for handling D channels. At each of these FS and FR points, tables are available to correctly route frames in both directions through the network. A more simplified representation of a chain from subscriber to PH is depicted in Figure 4.17. The FS and FR are each a point at which routing of D channel packets must be done, multiplexation of packets can be done, are layer 2 functions are performed either completely (FS) or partly (FR). At each of these points a table is available to store the relation between the different elements of the established path:

- the unique number of the D channel in the exchange
- the TEI
- the SAPI
- the addresses of the elements within the IRSU and IRIM
- the addresses of the Bd channels between the IRSU and local exchanges and between the local and transit exchanges
- the addresses of the Bd channels between the PH and the transit exchanges

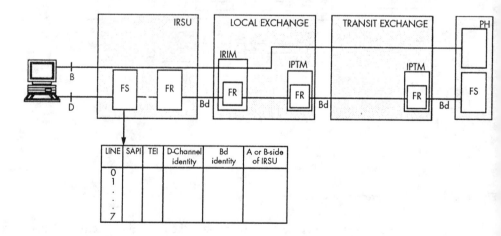

Figure 4.17 D channel routing.

The content of the table at each of the FR or FS functions is different. For example, at the FS function in the IRSU the relation has to be kept between D channel identity, line number identity, TEI, SAPI, the HDLC circuit in IPTM of the local exchange, and Bd channels between the IRSU and local exchanges and between the local and transit exchanges. With the help of the tables the D channel packets can be routed from the terminal to the PH and back even though there is not a single route available between the terminal and the PH — but a path with several points at which several channels are multiplexed into one.

4.3. OPERATION AND MAINTENANCE FOR DATA COMMUNICATION

By talking only briefly about operation and maintenance (O&M) aspects, we do not mean to create the impression that it is an unimportant matter. On the contrary, it is an area of high consequence that can influence the decision to go into one or another direction. As said before, it is our firm belief that the importance of O&M aspects has been overlooked up to now in choosing the right data communication means.

O&M is discussed here in the context of the minimum integration of packet-switching in ISDN. This means that B channels are not really treated by the ISDN; only the phase of setup of the connection to the PH is of interest. As demonstrated in previous sections, the D channel experiences several manipulations in the ISDN and is therefore well supervised and maintained by the network. However, for B as well as D channels a considerable part of the functionality related to the connection is located in the PSPDN. Thus, a complete supervision can be done only if the two parts (ISDN and PSPDN) are well coordinated.

Unfortunately, no standards exist today to line up the functions in the two networks. A complete O&M scheme is possible only by rationally planning tasks in two separate networks, which requires two different skills: ISDN and PSPDN skills. In what follows only the functions of the ISDN are discussed, but bear in mind that if the packet-switching function is fully integrated in the ISDN, no other functions are needed.

4.3.1. Errors and Alarms

All of the following discussion is related to D channel packet services only. The B channel services are supported by circuit mode connections to the PH, and thus all the functions of circuit-switching O&M apply. A Bb channel is handled as any other circuit mode call in the system. The PHI trunks are handled as all other 2 Mbit PCM systems. All indications that are related to an equipment failure, including software and firmware failures, are reported with the same alarm levels.

At the line side of the local exchange, the following is foreseen for D channels:

- Protocol errors detected at layers 1 and 2, are reported as alarms when they persist, but only for selected, specific subscriber terminations. Therefore a new classification is introduced: "sensitive lines." Transient errors are disregarded.

- Protocol errors made by the PH and detected by the local exchange are reported. A persistent failure to establish a link to the PH while the Bd continuity check is successful is considered a PH malfunctioning and will not be reported in the local exchange.

For the on–line supervision of layer 2, the following is foreseen:

1. global counters per circuit (15 counters per line), for a set of events two other counters are available per line (access):
   - retransmission counter
   - frame check sum (FCS) error counter
2. display of the link status (Q.921 state)
3. monitoring of the retransmission rate at the user–network interface

The quality of service experienced by a packet subscriber is defined at its subscriber-network interface. As such it is a layer 3 related item and thus controlled by the PH. The ISDN provides access only, and the standardized access procedures are defined in such a way that they do not deteriorate the service offered by the remote packet network. The functionality of the access network is minimal in order to distinguish between access problems and packet-network problems. Further QOS supervision in the frame handlers would make them more complex, which is not the purpose of the minimum integration approach.

- monitoring of the sequence numbering: no additional reporting or counting will be provided because of immediate protocol recovery
- error messages for system failures, generated and offered to local and central filtering

Excessive errors are reported by the frame-handling OBC based on errors and time period or total number of frames and dependant on the sensitive line category class mark. Further error filtering is done according to generic maintenance mechanisms.

At the outgoing trunk side of the local exchange, the IPTM module will report and raise alarms if its normal working is disturbed. The IPTM has an internal overload protection for itself, but it cannot protect the PH from being overloaded.

4.3.2. Display of Status

All configurable parameters for layers 1 and 2 can be displayed on the exchange maintenance terminal.

- On request: for subscriber and TEI, the link status for the user side of the FS can be displayed.

- Per line circuit: a display will list the active links.
- Per active datalink (layer 2 entity): a display is foreseen that will contain the actual status, information about the last link reset (reason and time), holding time and identity of the used Bd channel.

4.3.3. Traffic Measurement and Statistics

4.3.3.1. General

The framehandling functions are optimized for throughput. Elaborated statistics would decrease the maximum throughput. Therefore these statistics are better provided by the remote PH, because most functionality is concentrated there.

4.3.3.2. Network Aspects

The following counters are provided:

- counters per Bd channel on the PHI, keeping statistics about the CRC errors received, the frames received and sent, the number of times congestion is found, the discarded frames because of queue–overflow, and so on
- counters per internal Bd channel (between two FH modules) inside the local exchange
- percentage of CPU utilization of each frame handler module

4.3.3.3. User Interface

The following counters are provided:

- counters per D channel and per HDLC
- per (HDLC) (ISM, IRSU, and IPTM for PRA): compliance
- counters per link
- number of frames received and transmitted

4.3.4. Traffic Management

4.3.4.1. Failure in the PHI

In the case of an external failure, a report and alarm will be raised. At this point ISDN and PSPDN are interconnected. Manual intervention will be necessary to ensure that proper service is always available.

4.3.4.2. Frame Traffic Beyond Capacity

When the frame handler (as a whole: frame switch on the line board, frame relayer in the outgoing trunk (IPTM)) detects congestion in a certain direction, it will send RNR frames backward as soon this is appropriate. When necessary, it will modify RR into RNR. In case of processor overload, the processor will do the same. Under more heavy congestion or overload conditions, the frame handler will discard excessive frames from the queues. It is clear that proper dimensioning and resource management make keep congestion situations in the ISDN frame handling network very rare. The ISDN frame handler will normally have a higher excess capacity than the PH.

4.3.4.3. Processor Overload in the Frame Handler

Processor overload is limited by the X.25-layer 2-RNR mechanism. This effect will slow down the data transfer rate to an amount that can be handled by the system. (The PH is also protected by the same mechanism.)

Appendix A

Packet-Handling Background

This appendix provides historical and technical background on packet-handling techniques in general. For interested readers, it supplements the main text with further detail and discussion.

A.1. PACKET SWITCHING IN GENERAL

In the most limited sense, packet switching in the ISDN involves support of X.25 terminals connected by X.31 terminal adaptors to a narrowband ISDN access. In the most general sense, packet switching in the ISDN involves any transfer of information in the form of labeled blocks, with each block occupying the full available bandwidth during transmission. The label defines the routing of the block together with other control information. This conforms to the use of the word *packet* in such phrases as *fast packet switching*, *packetized speech*, and so on. The word *packet* is also used in CCITT Recommendation I.140 in this more general sense: transfer modes are classified as either "packet" or "circuit." CCITT Recommendation I.113 "Vocabulary of Terms for Broadband Aspects of ISDN" defines the following terms:

BLOCK   a unit of information consisting of a header and an information field
CELL   a BLOCK of fixed length identified by a label at layer 1 of the OSI model
FRAME a BLOCK of variable length identified by a label at layer 2 of the OSI model
PACKET an information BLOCK identified by a label at layer 3 of the OSI model

This is a fair attempt at pinning down the word *packet* provided that users of the word have the same idea about what the OSI model means in this generalized sense. Note that Q.931 messages and Signaling System Number 7 messages count as packets within this definition. However, the definition does not align with I.140, nor does it conform to informal usage (no one talks about "fast block switching" or "blockized speech"). In this book, therefore, as in CCITT and general usage, the meaning of *packet* depends on the context in which the word is used. Although the book concentrates on packet switching in the restricted X.25 sense, the more general sense justifies discussion of packetized signaling, additional packet mode (also called *new packet mode*), and asynchronous transfer mode. Because of its familiarity and importance, X.25 is used as the basis for discussion of more general packet handling.

We shall see that packetized signaling has most similarities to traditional packet switching as epitomized by X.25, whereas new packet mode and asynchronous transfer mode techniques progressively strip functionality and complexity, taking advantage of more modern transmission and switching technology to greatly extend the speed and applicability of the packet-switching technique.

A.2. X.25 PACKET HANDLING

X.25 and X.75 were defined with the background of data communication technology available in the early to mid 1970s. This basically consisted of the use of modems to transfer digital signals, as combinations of speech band frequencies, over analog links. The technique is still widespread and has a telephonic analogy in push-button sets and signaling systems as CCITT R2. The underlying analog nature of this technology is the opposite of the more modern ISDN concept, and the resulting low speed and often unreliable digital bitstreams influenced the definition of X.25 and X.75. Another major influence was the technology available for packet-switching, which could be performed only in software with limited and expensive processing and memory resources.

X.25 defines the user access to the network as an interface between user equipment (data termination equipment, or DTE) and network equipment (data communication equipment, or DCE). The DCE could be physically realized as a modem. X.75 defines the interface between packet-switching networks in a similar manner. Internal protocols within the network are undefined and are normally proprietary, so that PSPDNs are normally from a single supplier.

A.2.1. X.25 Level 1

This defines electrical characteristics of the DTE–DCE interface. It is based on the CCITT V–series of modems and basically provides for a point-to-point bitstream between user and network. This possibly unreliable duplex bitstream is the service offered to level 2.

A.2.2. X.25 Level 2

X.25 level 2 converts the bitstream of layer 1 into a controlled, high-quality, point–to–point packet stream (called a datalink) offered to level 3. It does this by delimiting blocks of bits called *frames* (some of which may carry packets), detecting and discarding errored frames, correcting errors by retransmission of frames containing packets, and controlling the flow of these frames. It is based on the high-level datalink control (HDLC) definitions of ISO, themselves based on the IBM synchronous data link control (SDLC) definitions. Because HDLC has many applications beyond X.25, we will discuss it here in some detail.

A.2.2.1. HDLC

HDLC operates on any bitstream and defines the following:

- delimitation of data blocks, called *frames*
- error detection in each frame
- labeling each frame with an address
- classification of frames as *command or response*
- frame types and related formats
- general procedures for use of the frame types

HDLC Delimitation

HDLC defines a FRAME (see Figure 2.5) as a string of bits with opening and closing flags (binary 01111110) delimiting the frame contents. Repetition of the flag within the contents is inhibited by "bit–stuffing," that is the insertion of a zero after any sequence of five ones, on transmission, and the removal of this zero on reception. Frames may be separated by further flags or an idle sequence of all ones.

HDLC Error Detection

Immediately before the final flag, HDLC defines a 32–bit or 16–bit frame check sequence (FCS) as a cyclic redundancy check on the preceding contents, which can be any number of bits. Incorrect frames are discarded on reception.

HDLC Address and C/R

Immediately after the initial flag, HDLC defines an address field of an integral number of octets. The last octet of the address has least significant bit one, whereas while preceding octets have least significant bit zero. HDLC defines frames as either *command* or *response* and includes a C/R bit for this purpose in the address. The C/R bit is normally used asymmetrically for each link direction, as a protection against inadvertent looping in transmission equipment (for example, a short circuit between send and receive directions). The remainder of the HDLC address is defined flexibly to allow for various link configurations, such as multidrop, with multiplexing of any number of frame streams onto a single bitstream.

HDLC Control

Following the address field, HDLC defines a control field of an integral number of octets containing both the frame type and control information such as sequence numbers. A wide range of frame types are defined.

HDLC Information

In certain frames, HDLC allows for the transfer of data by the user of the link. This data can be composed of any number of bits between the control field and the FCS. In the I.113 sense, this data is a packet.

To conclude this discussion of HDLC, we note that it is more a collection of possible frames and procedures applying to a variety of data communication applications than a specific interface specification. This generality may explain its success in the area of data communications. It is like a menu, from which a selection is made to define level 2 procedures for any particular application.

LAPB

The HDLC address capability is not used for level 2 of X.25, which is designed for a point–to–point configuration. Instead a 16–bit FCS is chosen. The C/R bit is made asymmetrical by choosing its significance to be opposite for the two directions (user–network and network–user). Two attempts were made to select frames and define procedures for level 2 of X.25 (called link access procedures). The first of these was called LAP: it was rather asymmetrical and used separate setup and reset procedures for each direction of data flow. It was replaced by the balanced procedures of LAPB (B = *balanced*). These provide for data link setup and disconnection, as well as the secure flow-controlled transfer of data on an established link. A typical sequence of LAPB frame exchanges is shown in Figure A.1.

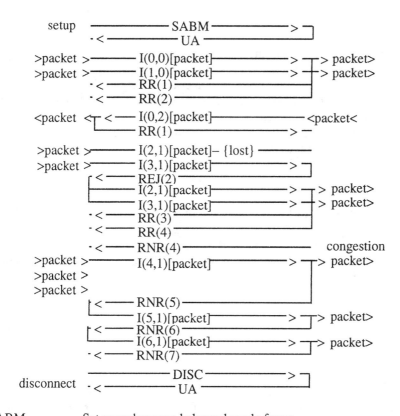

SABM	Set asynchronous balanced mode frame
DISC	Disconnect frame
UA	Unnumbered acknowledgment frame
I(s,r)[ ]	Numbered information frame:
	s: send sequence number
	r: receive sequence number
	[ ]: information field
RR	Receive ready supervisory frame
RNR	Receive not ready supervisory frame
REJ	Reject supervisory frame

Figure A.1. Typical LAPB frame exchanges.

LAPB Error Correction

Data is transferred in HDLC numbered information (I) frames identified by a send sequence number, incremented with each transmission. I frames are acknowledged by either backward I frames or supervisory frames containing a receive sequence number. frame loss is detected by sequence mismatch and is signaled with a reject (REJ) supervisory frame, which requests global retransmission (that is, all I frames starting with the lost one). A timer supervises acknowledgement of I frames and protects against los REJ or loss of the last I frame(s) of a sequence.

LAPB Flow Control

Flow control is achieved by limiting the number of I frames that may be transmitted without acknowledgment; this is called a *transmission window*. The window is normally rotated by means of the receive ready (RR) supervisory frame but may be reduced to one by means of the receive not ready (RNR) supervisory frame in cases of severe congestion.

LAPB Sequence Numbers

These may to be either modulo 8 or modulo 128. The latter is called *extended sequence numbering* and is signaled during link establishment.

Multilink

Optional multilink procedures are also defined in X.25. These run separate LAPB procedures on a group of load–shared links and define resequencing procedures to allow the packet stream to be randomly split over the link capacity on transmission, and recombined on reception.

A.2.3. X.25 Level 3

The relation of an X.25 packet to the bits at level 1 and the frames at level 2 is shown in Figure 2.6. X.25 level 3 uses the reliable stream of packets (in general, of any number of bits) provided by level 2 on each user link. It provides for switching packets containing user data from one user link to another.

Early versions of X.25 mention a possible connectionless service. This entails switching individual packets containing address and other information between users Such packets are called *datagrams*, because they are similar to letters contained in addressed envelopes switched by the postal service.

A.2.3.1. Virtual Circuits

X.25 level 3 now defines only a connection-oriented service, based on the concept of virtual circuits. Each user link is assigned a certain maximum number of simultaneous connections, identified by a label called *logical channel* present in every X.25 packet. Some packet layouts are shown in Figures 2.7 and 2.8. Establishment of a virtual circuit associates a logical channel La on a user link A to a logical channel Lb on a user link B. Clearing of a virtual circuit removes the association. For a permanent virtual circuit, establishment and clearing are done off-line by manipulating data in the network, under operator control. For a switched virtual circuit, establishment and clearing are done by a packet exchange on each user link. On an established virtual circuit, packets containing user data received on link A with logical channel La are transferred to link B, where the logical channel is changed to Lb, and vice versa. Figure A.2 shows some typical packet sequences for a switched virtual circuit. Note that the network is itself a black box, and its internal behavior is left for proprietary definition.

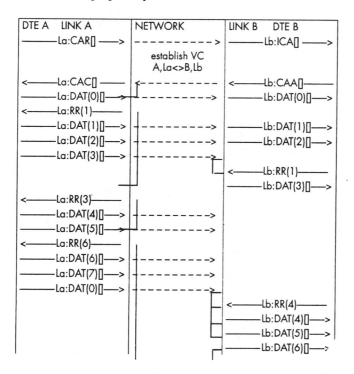

Figure A.2 Packet sequences for a switched virtual circuit.

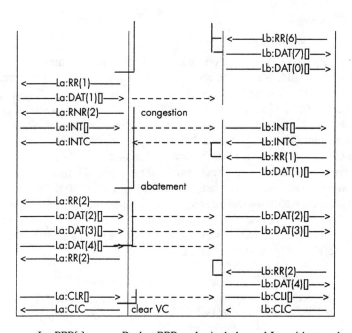

Lx:PPP[ ]	Packet PPP on logical channel Lx with user data [ ]
CAR	Call request packet
ICA	Incoming call packet
CAA	Call accepted packet
CAC	Call connected packet
CLR	Clear request packet
CLI	Clear indication packet
CLC	Clear confirmation packet
DAT(s)	Data packet with send sequence number s
RR(r)	Receive ready packet with receive sequence number r
RNR(r)	Receive not ready packet with receive sequence number r
INT	Interrupt packet
INTC	Interrupt confirmation packet

Figure A.2 (Continued)

Notes:
1. Sequence numbers modulo 8
2. No delivery confirmation
3. Window size 3
4. Maximum packet lengths equal at A and B

Figure A.2 (Continued)

A.2.3.2. Level 3 Flow Control

Flow of data is controlled as follows. For each virtual circuit on each link and in each direction, a window size is defined at establishment. The window size is the maximum number of data packets that can be sent without new authorization from the receiver. Each data packet has a send sequence number s. Authorization is sent as a receive sequence number r which allows transfer of data packets with r such that $s < r + w$, where w is the window size. The last r received by the sender is called the *lower window edge*, and updating r is called *rotating the window*. The window size may be reduced to zero by use of the RNR packet, for instance in cases of network congestion, and restored to its initial value by use of the RR packet.

Thus, on each direction of each virtual circuit, data packets are allowed from the source user by the network and from the network by the destination user. The network attempts to match these flows in an unspecified manner; it must provide sufficient flexibility to allow for transnetwork delays and jitter, but it must take care not to consume network buffering resources when the destination is accepting at a slower rate than the source can deliver. In the general case this in itself is not an easy problem, because the nature of the traffic can vary from intermittent transfers to continuous flow of data. It is exacerbated by considerations of processor, memory and link congestion within the network.

A.2.3.3. Packet Sizes

The standard maximum user data field length is 128 octets (this is often referred to as packet size, see our earlier comments about the word "packet"), but values of 16, 32, 64, 256, 512, 1024, 2048, and 4096, are allowed. The network must perform segmentation when a source sends a data packet that is too long for the destination; this is indicated by the more (M) bit, set in all but the last of the multiple Data packets delivered to the destination. In the opposite direction, the network delivers longer packets, formed by reassembling sequences of shorter packets with the M bit set in all but the last.

A.2.3.4. Delivery Confirmation

Normal data transfer, in which authorizations at source and sink are decoupled, may be overridden by a user requesting delivery confirmation in a data packet (with send sequence number s, say) by setting the Delivery (D) bit. This completely changes the significance of the receive sequence number in the backward direction: instead of a local authorization, it becomes an end–to–end acknowledgement, with the value $r = s + 1$ meaning that the user data in the data packet was delivered to the destination and was itself confirmed by a receive sequence number in a similar manner. When packet sizes vary between source and destination, interactions between D and M bits are interesting.

A.2.3.5. Qualifier

A further qualifier (Q) bit extends the user data by one bit. This has the effect only of complicating the network, which has to check that it takes the same value in all data packets of a segmentation sequence.

A.2.3.6. Interrupt

Flow control may be bypassed by a short interrupt packet containing user data. This is confirmed end to end by an interrupt confirmation packet. Only one such transfer is allowed in a given direction at any one time.

A.2.3.7. Reject

A rather bizarre packet retransmission option is supported by some X.25 networks. A user is allowed to send a Reject packet to request the network to retransmit packets that have already been acknowledged at level 2. If implemented, this requires buffering for retransmission at both level 3 and level 2.

A.2.3.8. Reset

Flow control may be reinitialized by reset procedures, using the reset request and reset confirmation packets. These have the effect of flushing any data being transferred on a virtual circuit and resetting all sequence numbers to zero. They are invoked in various error or problem situations and are also used to initialize a permanent virtual circuit.

A.2.3.9. Throughput Class

Throughput class is a measure of the steady state throughput that can be provided under optimal conditions in each direction of a virtual circuit. Some of its features require further study.

A.2.3.10. Negotiation

Throughput class, window sizes and maximum packet sizes may be negotiated during establishment, using the facility field in signaling packets. Various supplementary services (called *facilities*) also make use of this field. Because we are here concerned with packet handling, we will not discuss them further.

A.2.4. X.75

X.75 connects network to network rather than user to network. Its packet handling is essentially the same as X.25, with the multilink procedure often implemented for security and throughput reasons. Signaling is also similar, but extra signaling required between networks is contained in a utility field. For some obscure reason, the encoding for certain causes (reasons for release) is different from X.25.

A.2.5. X.25 Conclusions

A.2.5.1. *X.25 Importance and Influence*

We must first highlight the tremendous success, influence and importance of X.25 in the world of data communications. Following its definition, X.25 was rapidly accepted and formed the basis for PSPDNs implemented in many countries. X.25 capability became a must for communicating computers. As in this appendix, its structure, techniques, and terminology are an accepted and well-known starting point for general packet-switching discussions. Its LAPB specification, based on HDLC, formed the basis for further datalink protocols, such as LAPD (Q.921) and LAPM (V.42). Its three-level approach to networking was sanctified in the OSI model (which calls the levels *layers* and changes a few other names). Although other specifications defined at roughly the same time as X.25 use the same basic level structure (1 = bits, 2 = link, 3 = routing), the level 3 functions of X.25 define the OSI network service in CCITT Recommendation X.213. On this base is built the towering edifice of the OSI stack, the specification for open communications between different computing systems. The OSI model has influenced later versions of the X.25 recommendation. The word *level* has been replaced by layer, and certain other features have been included. Finally, its importance is underlined by the huge efforts expended in the ISDN to handle X.25 terminals and switch their packets.

A.2.5.2. *X.25 Reservations*

However, X.25 is a child of its time and cannot be frozen in stone as the one and only way of switching packetized information. More modern technologies require simpler and more basic standards, as we shall see. Even in the context of the late 1970s, one can criticize certain aspects of X.25, which are very difficult to implement efficiently:

- At level 2, the LAPB specification is not always rigorous. It requires interpretation, which may differ between implementations, and is sometimes rather internally inconsistent.
- At level 3, the in–band nature of the signaling, which is forced to be in strict sequence with data, makes it difficult to decouple data transfer functions from call control functions. This in–band characteristic is carried over to the OSI stack. In ISDN terms, we would say that it is difficult to separate the control plane from the user plane. The use of logical channels for connection identification means that it is possible for an incoming call and an outgoing call to collide. Other mechanisms (see packetized signaling) can avoid this and eliminate the need for special search procedures and the possibility of unnecessary call failure.
- At level 2 and level 3, the use of moduli 8 or 128 for sequence numbering complicates implementation. A general use of 128 at both levels would, in hindsight, have been preferable.

The window mechanism is suited to continuous transfers, but it is complicated for interactive or intermittent traffic, often resulting in four frame transfers for one user data item. The loss of a window update packet at level 2 can freeze data transfer indefinitely.

D bit and Q bits mix up end–to–end features with access features, complicating implementation of basic data transfer.

The huge range of packet sizes complicates network functions. Packet retransmission mixes up level 2 and level 3 functions and complicates buffering.

Reset is complicated, and in practice often results in user clearing.

It is not defined what to do with queued data when a virtual circuit is cleared.

Duplication of functionality between level 2 and 3 might also be criticized, in that both have sequence numbers and flow control, so that, say, some sort of retransmission at level 3 might remove the need for LAPB procedures. Alternatively, logical channels at level 2 might remove the need for level 3. However, the capability of level 2 to run one fine retransmission timer, and its ability to control the flow of all data on a link, were probably critical features at a time of high memory costs.

Finally, we must mention again the complete absence of an intranetwork specification. X.75 is appropriate between networks, but even in that case, discrepancies between national standards occasionally cause problems. Intranetwork protocols are basically proprietary and may or may not be based on an augmented X.75. Lack of standardization plagues attempts to integrate X.25 packet-switching into an ISDN of nodes made by different manufacturers. Such a situation would be absolutely unacceptable in the world of circuit switching.

We will now consider other forms of packet handling and compare them with the X.25 baseline.

A.3. PACKETIZED SIGNALING

A.3.1. Signaling System Number 7

A.3.1.1. Message Transfer Part

The message transfer part (MTP) of Signaling System Number 7 is discussed more fully in the main body of this book. We are here interested in comparison with X.25. Like X.25, the MTP has a three-level structure:
1. Level 1 defines a bitstream but allows for direct use of digital 64 kbps trunks between exchanges as well as modems.
2. Level 2 is similar in functionality to LAPB, converting the bitstream of layer 1 into a controlled, high-quality, point–to–point packet stream offered to level 3. The packets are called *messages*. However the manner in which this is done is very different. The MTP level 2 parts company with HDLC after taking over the delimitation and error detection functions. Frames are called *signaling units*.
3. Level 3 switches messages, just as X.25 level 3 switches packets. However, the messages are datagrams with a destination node in every message. The MTP network service is thus connectionless, with sequence guaranteed by use of the signaling link select. The MTP is suited to intermittent transfers, such as signaling, which is its main application. There is one maximum message size. Switching functions are thus much simpler than in X.25. To a certain extent, this simplicity reflects the fact that the MTP offers a network service to known users embedded in exchanges and other centers in the PSTN and ISDN. Its services are not open to the general public. Its users can be trusted to protect the underlying data communication network. On the other hand, level 2 and level 3 of the MTP have a great deal of network management functionality. Simplicity of switching, trust of users, and especially extensive network management are what one would expect of an intranetwork specification such as the MTP, in contrast with the extranetwork X.25 and X.75.

A.3.1.2. Signaling Connection Control Part

The signaling connection control part (SCCP) uses the underlying MTP network or networks, but extends its capabilities. First, in class 0 (unsequenced), and class 1 (sequenced) it provides for general connectionless datagram routing; this can be based on a network function identity (for example, a database) or, in the most general case, on an

E.164 number. Second, in class 2 (basic) and class 3 (flow controlled) it provides a connection–oriented service, again with generalized routing of the connection. Connection identification reflects the fact that SCCP operates over an underlying network rather a than point-to-point link. The logical channel approach of X.25 is replaced with the exchange of independently assigned local references for the connection; therefore, call collision cannot occur, unlike in X.25.

Class 2 provides essentially for a packet relay service, without flow control. It is designed for intermittent transactions such as user–to–user signaling. Segmentation and reassembly are defined.

Class 3 approaches closest to X.25. It extends class 2 with sequence numbers, window rotation, interrupt (called *expedited data*), and reset but does not include D bit or Q–bit. It therefore provides the OSI network service to its users.

However, SCCP users are still within the family of ISDN exchanges. Signaling for SCCP is therefore simple when compared with the rich variety of facilities available in X.25 and X.75. For coupling an SCCP connection to an external user connection, extra signaling is required. For user–to–user signaling associated to a circuit, this extra functionality is achieved automatically by the circuit–related ISUP signaling. For user–to–user signaling not associated to a circuit, this extra functionality may be achieved by embedding ISUP signaling in SCCP user data.

A.3.2. Digital Subscriber Signaling Number 1

DSS1 defines the interface between NB ISDN user and the ISDN. The D channel is used for packetized signaling and other purposes on either a point–to–point or point–to–multipoint subscriber access.

Layer 1 is much more than a simple bitstream. For our purposes we note that layer 1 contains D channel access procedures for a multipoint basic rate access (see Recommendation I.430) which rely on HDLC structuring of the D channel to provide for orderly access, and which rely on the fact that terminal equipment sends all ones when it has no frame to transmit. In addition, the supply of power on the basic rate access influences both layer 1 and its interaction with layer 2.

Layer 2 LAPD is based on LAPB of X.25. However, in addition to correcting certain inconsistencies in LAPB, LAPD's fuller specification of detailed procedures, use of modulo 128 at all times, and default values for system parameters allow it to make fuller use of HDLC:

- The address field is used more fully to identify both the function and the terminal using the datalink. This allows for multiplexing of various datalinks on the same D channel, which is necessary for support of a point–to–multipoint access.

- The unnumbered information (UI) HDLC frame is used, to manage terminal endpoint identifiers (TEIs) and to broadcast the layer 3 SETUP message on a point-to-multipoint access.
- The exchange identification (XID) HDLC frame is available in order to use nondefault layer 2 parameters on a datalink. (X.32 envisions the use of the XID as a means for preliminary identification and parameter negotiation after dialing in to an X.25 network.)

An established LAPD datalink may support an X.25 level 3, which is distinguished by the SAPI part of the HDLC address. From the point of view of X.25 layer 3, it provides a secure and flow-controlled packet stream.

An established LAPD datalink may also support a Q.931 layer 3, again distinguished by a SAPI value in the HDLC address. From the point of view of Q.931 it provides a secure and flow-controlled packet stream. In this case, the packets are called *Q.931 messages*. Q.931 also makes use of the unguaranteed UI service. Layer 3 Q.931 defines switching functions, but in a more general way than X.25. Like X.25, it provides for a connection, this time called a *signaling connection*. However, the use of logical channel is not appropriate on a point–to–multipoint access. Instead, a call reference is freely assigned by the call originator and used by both sides for the duration of the call. The use of a flag to distinguish messages to and from the call originator avoids the X.25 collision problem.

User data may be transferred on the signaling connection, both in signaling messages and in special USER INFORMATION messages. In this latter case, sequence numbers, windows, reset, interrupt, D bit, and Q bit are not supported, and a standard maximum length of user data is defined. The transfer of such messages is limited to a low rate by the network, and excess messages are discarded, with notification in a CONGESTION CONTROL message. This is also used like the X.25 RNR packet to stop flow in case of network congestion. The signaling connection is normally associated with a circuit–switched connection involving the use of one or more B channels. Identification of these, and the bearer capability that they are used to support, is an important function of Q.931.

When compared with X.25, call establishment in Q.931 is more complicated:

- Provision is made for overlap transmission of the called address, meaning that it may be sent one digit at a time (for instance, from a telephone).
- A local acknowledgment is returned to the call setting (SETUP) message and to the last address signal if overlap is used. This is necessary for setup broadcast and to inform a terminal of call progress.
- Responsibility for alerting the called party (ringing) is passed to the caller.
- The connection completion (CONNECT) message from the called is locally acknowledged. This is used to award the call after SETUP broadcast, but it also

prevents the called from sending messages immediately after the CONNECT, in contrast to X.25.

Release is also more complicated. Three messages, rather than two, are used if there is a circuit–switched connection involved. In addition, Q.931 uses a status message to inform of both nonfatal errors and the present perceived call state, and it provides for notification of certain events (such as hold and suspend) during the call. The Q.931 messages themselves use a generalized tag–length–value encoding for information elements, in many ways similar to that used for facilities in X.25. (*Tag* refers to identification of the information meaning; *value* refers to the information contents of *length* octets.) However, the encoding also applies to addresses, which are treated with more privacy than in X.25. Supplementary services may be invoked in a manner similar to X.25 facilities, but the use of keypad characters is also possible for dumb terminals.

Most of the extra complexity in Q.931 arises both from the more complex access arrangements and from the more extensive range of terminals and call handling possibilities supported in the ISDN.

A.4. LAPM

Another LAP is worth mention here, namely LAPM, used between V.42 modems to improve the link quality. Like LAPB and LAPD, it uses the ubiquitous HDLC. As in LAPD, the HDLC address flexibility is used to identify various streams of frames, the UI is used for management, and the XID is used to negotiate parameters. However, there is no layer 3 in the normal sense; the "packets" are simply groups of characters assembled and disassembled on a local V.24 interface, so that the modem appears of higher than normal quality. Both 16–bit and 32–bit FCS are possible, and the efficiency of data transfer is improved by use of the selective reject (SREJ) frame, this allows for retransmission of just one errored I–frame. LAPD and LAPB use global retransmission, in which all I–frames starting with the errored frame are retransmitted.

LAPM illustrates both the flexibility of HDLC and the influence of LAPB in other areas of data communication.

A.5. NEW PACKET MODE (ADDITIONAL PACKET MODE BEARER SERVICE)

In the early 1980s, as standardization work proceeded for the narrowband ISDN, the position of X.25 as "the" packet-switching definition was reexamined, particularly in the United States, where X.25 had never gained the same acceptance as in Europe.

In the United States, X.25 never really displaced connectionless datagram switching. The ARPANET datagram network was of vital importance in the history of packet-switching. It preceded X.25 and was the source for much seminal material in the nature and behavior of packet-switching networks and in the application of queuing theory to such systems. The ARPANET protocols, with an end-to-end connection–oriented protocol (transmission control protocol, or TCP) making use of an underlying connectionless protocol (internet protocol, or IP), formed the basis for the United States Department of Defense TCP/IP standard, which is popular for local area networking. In addition, the TCP was influential in defining 4 of the Transport Layer in the OSI stack (X.224), with its general procedures for end–to–end resequencing and retransmission in case of loss.

This predilection for the simplicity of a datagram service is also evidenced by the transaction capabilities application part (TCAP) of Signaling System Number 7, which uses the connectionless SCCP and also had its origin in the United States. American experts were therefore often rather skeptical about the merits of X.25, even while it was under definition. Some of their objections are mentioned in A.2.5 above, and they were always concerned about its complexity and the consequent cost and efficiency of implementation. When confronted with the additional complexity of X.31 for support of X.25 terminals on an ISDN access, their tendency was therefore to ask if a simpler approach was not possible to the basic problem of getting packetized data from one ISDN subscriber to another. The answer was yes, and we will now go through the basic arguments.

1. LAPD uses the HDLC address to multiplex various frame streams onto a single D channel.
2. Each stream is identified by its nature (SAPI) and endpoint (TEI), and a terminal can have multiple TEIs for one SAPI.
3. For signaling and X.25 (distinguished by SAPI) streams run from terminals to entities in the network, and an LAPD is operated between network and terminal.
4. However, other streams can run directly between terminals, which might run LAPD or other procedures end to end.

Point 4 is the essence of frame relay, in which frames, not packets, are switched within the network. Note that it is a connection–oriented approach, with the TEI identifying the connection at each access. It is extremely attractive in comparison with X.25 and X.31 procedures for the following reasons:

1. The signaling to establish and release datalinks can be performed outband (SAPI = 0) and can be integrated in the ISDN by adapting Q.931.
2. Layer 3 is not required on the connection.
3. A terminal multiplexes calls at layer 2 and need only implement LAPD or another protocol on an end–to–end basis.

4. The network need only recognize the HDLC address and relay on this basis, allowing for much simpler processing and much higher throughputs.

Recommendation I.122 distinguishes between frame relay 1, in which the end–to–end protocol is undefined, and frame relay 2, in which the end–to–end protocol uses LAPD frames, a fact which might be made use of by the network in certain cases. I.122 refers to "core" LAPD functions handled by the network. These are essentially yet another selection from the HDLC menu, namely HDLC delimitation, HDLC error detection, and HDLC address called data link connection identity (DLCI), together with limitation to a maximum number of integral octets per frame, and some sort of congestion control.

For frame relay 2, HDLC control and information functions run end to end. The only disadvantages with frame relay are that if the loss rate on the accesses becomes high, the end–to–end retransmission becomes inefficient, and that network flow and congestion control rely on simple mechanisms such as congestion notification and frame discard and user cooperation with these mechanisms. If these are felt to be problematic, then LAPD procedures with local retransmission and windows may be introduced on the network boundaries. This variant of new packet mode is called *frame switching*. Frame switching increases the complexity of processing in the network, but it still reduces the switching to one level rather than two and is therefore simpler than X.25. As one might expect, this variant is more popular in Europe.

The new packet mode approach of frame relay or frame switching obviously can be extended to the B channel, or to groups of B channels. Its simplicity allows for bit rates up to 2 Mbps to be handled within the network, and there is talk of incredible (in X.25 terms) speeds of 45 Mbps for frame relay. As for X.25, the intranetwork treatment for new packet mode is not yet defined. Moreover, at the time of this writing, CCITT standardization of the required signaling is not yet stable. This has not prevented the successful introduction of frame relay networks in the United States. In Europe, the tendency has been to stick to X.25, although frame relay and switching concepts are used in the ETSI packet handler interface (PHI) to concentrate D channel X.25 traffic onto multiplexed B channels called *Bd channels*.

New packet mode may be seen as the final evolution of the HDLC family of packet-handling techniques. As we have seen, this evolution has involved the gradual stripping of functionality until, for frame relay, only HDLC delimitation, error detection, and address (for routing) remain significant for transfer of packetized data.

The extension of the packet-switching technique to the transfer of more general information than computer data, at speeds of 150 or even 600 Mbps, requires a critical examination of even this minimal HDLC functionality, as we will now see.

A.6. ASYNCHRONOUS TRANSFER MODE (ATM)

A.6.1. ATM and HDLC

Both HDLC and ATM divide up bitstreams into blocks. We will now consider their differences and why HDLC is inappropriate for broadband ISDN.

A.6.1.1. Block Length and Idle

ATM uses a fixed block length with idle blocks between active blocks, whereas HDLC allows variable block length with idle octets or bit sequences between blocks. ATM is therefore a great deal simpler than HDLC. Once cell delineation has been established, processing can proceed in a periodic manner at the basic cell frequency. In contrast, HDLC processing must reinitialize after each frame and must examine the received bit stream in a nonperiodic manner. For the same level of technology, ATM is therefore capable of higher speeds, at the cost of flexibility.

Note that the use of fixed-length frames as an HDLC simplification is not possible because of its bit–stuffing technique. Note also that fast packet-switching techniques for speech also opt for fixed block lengths, as discussed earlier.

A.6.1.2. Delineation and Delimitation

ATM take advantage of its fixed-length block by defining an elegant technique for cell delineation, sharing the delineation overhead with error detection and correction. HDLC uses extra bandwidth to delimit its frames with flags and to stuff zeros to avoid flag imitation. Bit stuffing and removal complicates termination equipment, again reducing maximum attainable speed at the same level of technology.

Thus ATM, besides giving periodicity in block handling, also gives periodicity in memory accesses for transmission and reception, whereas HDLC memory access is more random. This allows again for simpler design and higher speeds.

A.6.1.3. Error Detection

ATM protects only the header (not the user information) with a code capable of correcting single-bit errors and does not discard a cell with a corrected header. HDLC protects the whole block with a code capable of detecting an error and must discard an errored frame. HDLC is thus designed for computer data, in which a single-bit error renders the whole frame invalid. Any error must be corrected by retransmission. Bit stuffing reduces the power of error detection. ATM is designed for more general information flow. For instance, a bit error in a group of speech or video samples does not necessarily render the whole group invalid. In addition, even for computer data, the users may employ such techniques as forward error correction, not possible via HDLC links.

A.6.1.4. Routing and Address

There is little to choose from here; the fixed ATM routing label could be implemented as an HDLC address.

A.6.1.5. Control

ATM defines only those control fields of concern to cell switching, and cells may contain any kind of information. Definition of application–related features is left to the ATM adaptation layer, which operates over the network and converts cell streams into bit streams (with smoothing of delay jitter) or longer packets (with segmentation and reassembly).

The set of HDLC frame types, used as elements of protocols such as LAPB and LAPD, is dedicated to transfer of computer data. However, there is no reason that these types and related procedures should not be used on top of an adaptation layer.

A.6.2. ATM and New Packet Mode

A.6.2.1. Frame Relay

The frame relay technique is very similar to ATM in its simple functionality. The main difference lies in the use of HDLC rather than on the ATM functions discussed here above, which allow us to conclude the following:

1. Frame relay is not capable of the same switching throughput as ATM for the same level of technology. At the same throughput it requires more complex switching equipment and is often handled by high-performance software rather than hardware.
2. Frame relay is designed only for transfer of information for which the loss of a single bit is disastrous. It will not, therefore, support forward error correction, and it is inappropriate for more general information flows than computer data.
3. Frame relay is less resilient against bit errors.
4. Frame relay is therefore aimed at a particular group of applications in the ISDN, whereas ATM aims to be a comprehensive technique for general applications at speeds up to 600 Mbps.

However, the great functional similarity means that an ATM network would be admirably suited to support of a frame relay service, as evidenced by the Draft CCITT

Recommendation I.5xz on interworking for the frame mode bearer service (another name for new packet mode). In addition, the wide availability of HDLC chips and the numerous applications of these at speeds up to 2 Mbps suggest that frame relay is an important step on the road to ATM and that adaptation of HDLC to ATM streams will be an important component of the evolving ISDN.

A.6.2.2. Frame Switching

The use of LAPD or other procedures for retransmission and flow control would be completely inappropriate on the boundaries of an ATM network. Even more so than frame relay, frame switching would be wrong for services such as speech or video: besides complicating the system and decreasing throughput, it is obvious that a retransmitted or flow-controlled (delayed) audio or video sample is useless. However, as for frame relay, the ATM network would form a suitable backbone for support of the more special-purpose frame switching access.

A.6.3. ATM and X.25

This is obviously a rather spurious comparison, but we conclude this appendix with a mention to show how far packet-switching has changed since the definition of X.25:

Layers:	ATM has one layer; X.25 has three.
Link:	see the comparison with HDLC above.
Sequence numbers:	ATM has none; X.25 has them at two levels.
Flow control:	ATM monitors streams and discards excess cells; X.25 uses two levels of window and RNR.
Lengths:	ATM has one (48 octets of information); X.25 has any number of bits from 16 to 4096 octets of user data.
Others:	ATM has no delivery confirmation, no interrupt, no reject, no reset; X.25 has all of these features.
Switching:	ATM switches in hardware; X.25 requires software.
Signaling:	ATM is outband; X.25 is inband.
Throughputs:	ATM utilizes speed up to 600 Mbps; X.25 utilizes speeds up to 64 kbps.

ATM thus aims to switch packets 10,000 times more quickly than X.25.

Abbreviations

ACM	Address Complete Message
AK	Acknowledgment
ANM	Answer Message
APMBS	Additional Packet Mode Bearer Service
ATM	Asynchronous Transfer Mode
BA	Basic Access
BC	Bearer Capability
BECN	Backward Notification
CAN	Customer Access Networks
CAPS	Calls Per Second
CNIU	Customer Network Interface Unit
CC	Country Code
CD	Count Down
CGW	Customer Gateway
CIC	Circuit Identification Code
CPE	Customer Premises Equipment
DCE	Data Circuit Terminating Equipment
DDI	Direct Dialing In
DLC	Data Link Connection
DLCI	Data Link Connection Identifier
DN	Directory Number
DNIC	Data Network Identification Code
DQDB	Distributed Queue Dual Bus
DT2	Dataform 2
DTE	Data Terminal Equipment
DTIC	Digital Trunk Interface Circuit
DTM	Digital Trunk Module
ECN	Explicit Congestion Notification
EGW	Edge Gateway
FAR	Facility Request
FCE	Front-end Control Element

FCS	Frame Check Sequence
FDDI	Fiber Distributed Data Interface
FDLCI	frame datalink identifier
FECN	Forward Direction
FISUs	Fill-in Signal Units
FMBS	Frame Mode Bearer Services
FR	Frame Techniques
FRBS	Frame Relaying Bearer Service
FS	Frame-Switching
HDLC	High-level Data Link Control
HEC	Header Error Control
HLC	High-Layer Compatibility
IAM	Initial Address Message
IC	Interface Circuit
ICON	ISDN Concentrator
ILC	ISDN link controller
IN	Intelligent Networks
IPTM	ISDN Packet Trunk Module
IPTMu	ISPBX Interface Module
IPTMn	Network Interface
IRIM	ISDN Remote Interface Module
IRSU	ISDN Remote Subscriber Unit
ISCTXs	ISDN Centrex Groups
ISDN	Integrated Services Digital Network
ISM	ISDN Subscriber Module
ISPBX	ISDN private exchange
ISSI	Interswitching System Interface
ITM	ISDN Trunk Module
LANs	Local Area Networks
LAPB	Link Access Protocol Balanced
LAPD	link access protocol D
LCI	Logical Channel Identifiers
LCN	Logical Channel Number
LSSUs	Link Status Signal Unit
LT	Line Termination
MAC	Medium Access Control
MAN	Metropolitan Area Network
MCU	Module Control Unit
MID	Message Identifier
MSN	Multiple Subscriber Number

MSS	MAN Switching Systems
MTP	Message Transfer Part
NDC	National Destination Code
NPM	New Packet Mode
MSUs	Message Signal Units
NT	Network Termination
NTN	National Terminal Number
OBC	On–Board Controller
OBCI	On–Board Controller Interface
O&M	Operation and Maintenance
PA	Permanently Assigned
PABX	Private Automatic Branch Exchange
PBAs	Printed Board Assemblies
PH	Packet Handler
PHI	Packet Handler Interface
PLL	Permanent Logical Link
PLLC	Permanent Logical Link Connections
PMBS	Packet-Mode Bearer Services
PPU	Packet Protection Unit
PRA	Primary Rate Access
PSM	Packet-switching Module
PSPDN	Packet-Switched Public Data Network
PSPDNs	Packet-Switched Public Data Networks
PVC	Permanent Virtual Circuit
QA	Queued Arbitrated
REJ	Reject
REL	Release Message
RETMO	Retiming and Monitor Channel Handling Unit
RLC	Release Complete Message
RR	Receive Ready
RNR	Receive Not Ready
SAPI	Service Access Point Identifier
SAR	Segmenting And Reassembly
SCCP	Signaling Connection Control Part
SDH	Synchronous Digital Hierarchy
SDLC	Synchronous Data Link Control
SLS	Signaling Link Selection
SMDS	Switched Multimegabit Data Service
SN	Subscriber Number
SNA	System Network Architecture

SNI	Subscriber Network Interface
SSF	Subservice Field
SSN	Subsystem Number
ST	Segment Type
STE	Signaling Terminal
STP	Signaling Transfer Point
TCE	Terminal Control Element
TCAP	Transaction Capabilities Application Part
TCP	Transmission Control Protocol
TEI	Terminal Endpoint Identifier
TMN	Telecommunication Management Network
TEI	Terminal Endpoint Identifier
UIC	U Interface Transceiver Chip
UNI	User-to-Network Interface
USBS	User Signaling Bearer Service
USR	User Information Message
UUS	User-to-User Signaling
VAP	Videotex Access Point
VC	Virtual Call
VCI	Virtual Channel Identifier
VPI	Virtual Path Identifier
WAN	Wide Area Network

INDEX

Access channel negotiation, 116
Access charge, 37
Access fee, 37
Activation, 67
Address complete message, 104
ADPCM, 146
Alert, 86
Answer message, 104
Associated approach, 95
ATM, 53
Attendant positions, 92
Attributes in the U-plane, 114

B channel, 13
Basic connection-oriented class, 100
Basic connectionless class, 100
BD channel establishment, 73
Bd channel establishment, 73
Bearer services, 25
Bearer service negotiation, 116
Bit-stuffing technique, 195
Block, 177
Block length, 195
Bridges, 141
Buffering techniques, 44
Bursty traffic, 36

Call clearing, 116
Call control, 108
Call control mapping, 127
Call setup, 83, 108
Call setup delay, 35, 80
Cell, 177
Cell relaying, 22
Cell structure, 147
Centrex, 79
Charging, 75
Charging principles, 37
Circuit switching, 14, 75
Common channel signaling, 84
Congestion, 43
Congestion avoidance procedures, 119
Congestion control, 26, 43, 86, 118, 194
Congestion recovery procedures, 119
Connect, 86
Connection confirmation message, 103
Connection oriented, 93
Connection request message, 103
Connectionless, 93, 129
Connectionless datagram, 189
Connectionless packet switching, 52, 138
Constant bit rate services, 147
Control congestion, 109
Control plane, 109
Core functions, 114
Crossconnect, 141
Customer access networks, 136
Customer gateway, 137
Customer network inteface unit, 137
Customer premises equipment, 137

D bit, 188
D channel, 14, 26
D channel protocol unit, 161
Data link connection, 109
Data link control, 114
Datalink, 26
Datalink connection identifier, 109

Datalink layer, 60
Datalinks, 62, 64
Datagram type, 129
Deactivation, 67
Defense mechanisms, 118
Delivery confirmation, 185
Dialogue oriented, 36
Digital trunk interface circuit, 155
Dimensioning, 43, 118
Directory enquiries, 92
Directory number, 30
Disconnect, 86
Distributed control, 44, 78
Distributed queue dual bus, 129
Division of revenue, 80
DLCI, 74
DLCI negotiation, 116
DNIC, 31
Double call control, 108
Double subscriptions, 76
Double-charging, 76
DQDB, 52
DQDB protocol stack, 134
DSS1, 190
DTM, 54
Dual bus network, 129
Dynamic window, 118

E.164, 31, 75
Echo control, 150
Edge gateway, 137
Exchange termination, 160
Explicit congestion notification, 121
Extensions, 76

Facilities fee, 38
Fault localization, 80
FDDI, 130
FISUs, 96

Fixed-length frames, 195
Flag, 96
Flow control, 26, 40, 109
Flow control connection-oriented class, 100
Forced idleness, 118
Frame, 17, 177
Frame acquisition, 145
Frame alignment, 145
Frame delay, 118
Frame discard, 121
Frame loss, 118
Frame relaying, 21, 52, 62, 83, 109
Frame relaying resources, 118
Frame switching, 21, 52, 62, 83, 109
Fully integrated network, 6

Gateways, 141
Global title, 101
Granularity, 118

HDLC, 85, 179
HDLC address and C/R, 179
HDLC address called data link connection identity, 194
HDLC control, 180
HDLC delimitation, 179, 194
HDLC error detection, 179, 194
HDLC framing, 145
HDLC information, 180
Header error control, 149
High layer compatibility, 29
Hybrid private networks, 92

ICON, 54
In band, 17, 49
In-band signaling, 92
Incoming calls, 73

Indicator bits, 96
Information elements, 116
Information transfer capabilities, 75
Initial address message, 104
Interactive database access, 36
Interactive text messages, 92
Intermediate storage, 118
Internal datagram, 93
Interrupt, 186
Interworking, 125
Intranetwork protocol, 82
IPTM, 65
ISDN concentrator, 39
ISDN link controller, 154
ISDN module for remote access, 155
ISDN multiplexer, 40
ISDN private exchange, 155
ISDN subscriber module, 155
ISDN user part, 104
ISM, 54, 155
ITM, 54

LANs, 52
LAPB, 26, 27, 29, 64, 180
LAPB error correction, 182
LAPB flow control, 182
LAPB sequence numbers, 182
LAPD, 26, 29, 64, 85
LAPM, 192
Layer 1, 16, 58, 67
Layer 2, 16, 58, 67, 85
Layer 3, 16, 86
Length indicator, 96
Level 1, 94
Level 2, 93, 96
Level 3, 93, 98
Level 3 flow control, 185
Level 4, 93
Levels of congestion, 118
Line termination, 162
Logical channel identifiers, 60

Loss rate, 194
Low-speed data applications, 92
LSSUs, 96

MAN, 22, 52
Material cost, 78
Maximum integration, 8, 10, 32, 50, 75
Medium access control, 130
Memorandum of understanding, 62
Message transfer part, 93, 189
Metropolitan area network, 129
Mild congestion state, 118
Minimum integration, 8, 62, 50
MSUs, 96
Multilink, 182
Multiplexation, 62
Multiplexing, 62, 65

Negotiation, 187
Negotiation of parameters, 116
Network actions, 122
Network congestion, 122
Network resources, 121
Network structure, 99
New packet mode, 52, 83
NTN, 31
Numbering plan, 30, 31, 39, 75
Numbering system, 30

On demand, 62
On-board controller, 161
Operation and maintenance, 76
Out-of-band, 17, 52, 53, 82
Out-of-band signaling, 92
Outgoing calls, 73
Overload capacity, 119

Overload control, 43

Packet, 17, 177
Packet delay, 35
Packet handler, 5, 14, 28, 50
Packet handler interface, 73, 76
Packet relaying, 20
Packet sizes, 185
Packet switching, 14, 20
Packet-mode bearer services, 66
Packetized data messages, 83
Packetized messages, 83
Packetized signaling, 178
Packetized speech, 177
Packetizing speech, 146
Packets, 4
Peer function, 85
Performance degradation, 118
Performance figures, 78
Performance requirements, 157
Permanent logical link connections, 109
Permanent virtual circuit, 66, 67
Physical layer, 60
Processing capacity, 118
Provision withdrawal, 69
PSM, 28, 54, 56
PSPDN, 4, 49
Pulse code modulation, 146

Q3 network layer protocol, 139
Q.921, 26, 32, 85
Q.931, 27, 32, 86
Q bit, 188
Qualifier, 186
Quality of service, 80, 121
Quality of service degradation, 121

Receive sequence number, 41
Reject, 186
Release, 86
Release complete, 86
Release complete message, 104
Release message, 104
Reliability, 78
Reset, 186
Router network elements, 136
Routing header, 149

S interface circuit, 155
SAPI, 27, 65, 85
SCCP, 93
Semipermanent, 62
Send sequence number, 40
Send window, 41
Sequence number, 96
Sequenced connectionless class, 100
Service 1, 88
Service 2, 88
Service 3, 88
Setup, 86
Severe congestion region, 119
Short-message service, 91
Signaling connection control part, 100, 189
Signaling link select, 98
Signaling link selection, 101
Signaling messages, 83
Signaling point, 99
Signaling transfer point, 95
Silence recognition, 146
Smart phones, 91
SMDS, 52
Speed of call, 35
Subaddress, 29
Subscriber Numbering, 29
subscription, 66
Subscription profiles, 81
Subservice field, 98
Supplementary services, 83

Supplementary services, 25
Switched multi-megabit data services, 129
Synchronisation pattern, 145
Synchronous framing, 145

TEI, 26, 64, 85
Teleaction, 92
Telemetry, 92
Teleservices, 25, 29, 122
Terminal control element, 159
Terminal function, 85
Throughput, 118
Throughput class, 186
Transmission links, 118

Usage fee, 37
User information, 86
User information message, 104
User plane, 109
User service profile, 66
User signaling bearer service, 89
User signaling Service 4, 83
User-network signaling, 85
User-to-user network interface, 109
User-to-user signaling, 11, 84, 190

V11, 56
V28, 56
Videotex, 124
Virtual call, 66, 67
Virtual channel identifier, 149
Virtual circuits, 183
Virtual connection, 36
Virtual path identifier, 149

Wide area network, 129
window size, 26

X.121, 31, 75
X.25, 16, 49
X.31, 32, 50
X.75, 49, 81

Acknowledgments

A book about a complex subject like telecommunications cannot be written by one person alone.

While preparing my manuscript, I had the sizeable advantage of being able to build upon the considerable expertise available within my company, Alcatel Bell, which was kind enough to allow me to use this material.

More particularly, I am grateful to the following persons for their highly respected contributions.

Guido Adams, who originated the article that was presented during the ISS in Stockholm in May of 1990, which was the basis for this book.

Tony Donegan for his valuable input of CCITT N7 and the comparison of X.25 with other signaling types.

Daniel Deloddere for this contribution to my discussion of MAN technology.

Equally important was the graphical support that I got from:

Linda Bleyenberg, who did the page layout and who made the necessary corrections to the figures.

My wife, Lisette, who not only typed my manuscript but who—more importantly for me—gave me moral support during the many evenings and weekend hours spent preparing this book.

The Artech House Telecommunications Library

Vinton G. Cerf, *Series Editor*

Advances in Computer Systems Security, Rein Turn, ed.

Advances in Fiber Optics Communications, Henry F. Taylor, ed.

A Bibliography of Telecommunications and Socio-Economic Development by Heather E. Hudson

Codes for Error Control and Synchronization by Djimitri Wiggert

Communication Satellites in the Geostationary Orbit by Donald M. Jansky and Michel C. Jeruchim

Current Advances in LANs, MANs, and ISDN, B.G. Kim, ed.

Design and Prospects for the ISDN by G. Dicenet

Digital Cellular Radio by George Calhoun

Digital Signal Processing by Murat Kunt

Digital Switching Control Architectures by Giuseppe Fantauzzi

Digital Transmission Design and Jitter Analysis by Yoshitaka Takasaki

Disaster Recovery Planning for Telecommunications by Leo A. Wrobel

E-Mail by Stephen A. Caswell

Expert Systems Applications in Integrated Network Management, E.C. Ericson, L.T. Ericson, and D. Minoli, eds.

A Guide to Fractional T1 by James Trulove

Handbook of Satellite Telecommunications and Broadcasting, L. Ya. Kantor, ed.

Information Superhighways: The Economics of Advanced Public Communication Networks by Bruce Egan

Integrated Broadband Networks by Amit Bhargava

Integrated Services Digital Networks by Anthony M. Rutkowski

International Telecommunications Management by Bruce R. Elbert

International Telecommunication Standards Organizations by Andrew Macpherson

Introduction to Satellite Communication by Bruce R. Elbert

Introduction to Telecommunication Electronics by A. Michael Noll

Introduction to Telephones and Telephone Systems, Second Edition by A. Michael Noll

The ITU in a Changing World by George A. Codding, Jr. and Anthony M. Rutkowski

Jitter in Digital Transmission Systems by Patrick R. Trischitta and Eve L. Varma

LANs to WANs: Network Management in the 1990s by Nathan J. Muller and Robert P. Davidson

The Law and Regulation of International Space Communication by Harold M. White, Jr. and Rita Lauria White

Long Distance Services: A Buyer's Guide by Daniel D. Briere

Mathematical Methods of Information Transmission by K. Arbenz and J.C. Martin

Measurement of Optical Fibers and Devices by G. Cancellieri and U. Ravaioli

Meteor Burst Communication by Jacob Z. Schanker

Minimum Risk Strategy for Acquiring Communications Equipment and Services by Nathan J. Muller

Mobile Information Systems by John Walker

Networking Strategies for Information Technology by Bruce Elbert

Optical Fiber Transmission Systems by Siegried Geckeler

The PP and QUIPU Implementation of X.400 and X.500 by Stephen Kille

Principles of Signals and Systems: Deterministic Signals by B. Picinbono

Private Telecommunication Networks by Bruce Elbert

Radiodetermination Satellite Services and Standards by Martin Rothblatt

Residential Fiber Optic Networks: An Engineering and Economic Analysis by David Reed

Setting Global Telecommunication Standards: The Stakes, The Players, and The Process by Gerd Wallenstein

The Telecommunications Deregulation Sourcebook, Stuart N. Brotman, ed.

Telecommunication Systems by Pierre-Girard Fontolliet

Television Technology: Fundamentals and Future Prospects by A. Michael Noll

Telecommunications Technology Handbook by Daniel Minoli

Telephone Company and Cable Television Competition by Brotman

Terrestrial Digital Microwave Communications, Ferdo Ivanek, ed.

Troposcatter Radio Links by G. Roda
Voice Processing by Walt Tetschner
Voice Teletraffic System Engineering by James R. Boucher